International Conference on Advanced Science and Technology (ICAST 2020)

Held online due to COVID-19

IOP Conference Series: Materials Science and Engineering
Volume 1071

Jakarta, Indonesia
28 November 2020

ISBN: 978-1-7138-3345-1
ISSN: 1757-8981

Printed from e-media with permission by:

Curran Associates, Inc.
57 Morehouse Lane
Red Hook, NY 12571

Some format issues inherent in the e-media version may also appear in this print version.

Institute of Physics
Dirac House, Temple Back
Bristol BS1 6BE UK

Phone: 44 1 17 929 7481
Fax: 44 1 17 920 0979

techtracking@iop.org

Additional copies of this publication are available from:

Curran Associates, Inc.
57 Morehouse Lane
Red Hook, NY 12571 USA
Phone: 845-758-0400
Fax: 845-758-2633
Email: curran@proceedings.com
Web: www.proceedings.com

TABLE OF CONTENTS

CIVIL & MECHANICAL ENGINEERING

MATERIAL SCIENCE & MECHANICAL ENGINEERING

INFORMATION ENGINEERING & TECHNOLOGY

IOP Conf. Series: Materials Science and Engineering 1071 (2021) 011001 doi:10.1088/1757-899X/1071/1/011001

Preface

Distinguished guests and participants,
Assalamualaikum Warahmatullahi Wabarakatuh.
Good Morning Ladies and Gentlemen.
Welcome to International Conference on Advanced Science & Technology (ICAST) 2020

On behalf of the Organizing Committees, we would like to express a warm welcome to all delegates and participants to the ICAST International Conference on Science and Technology 2020. It is a great honor for all collaboration institution including universiti teknikal malaysia melaka, university of bung karno, university of HKBP Nommensen and Sada Utama Indonesia to support the conference.

This conference is an important forum for exchange of information and research results among us, who come from different countries, different educational and research institutes, and different research interest. The main theme of conference is Innovation Research for Science and Technology in Industry 4.0 has been chosen. ICAST 2020 aims to bring together researchers, scientists, engineers, and scholar students to exchange and share their experiences, new ideas, and research results about all aspects of science, engineering and Technology, and discuss the practical challenges encountered and the solutions adopted. The conference invites delegates from across Indonesian and South East Asian region and beyond, and will be attended by participants from abroad university academics, researchers, practitioners, and professionals across a wide range of industries. Due to Covid-19 outbreak, the conference 1st International conference on advanced science and technology has been successfully held virtual on November 28th, 2020. Total more than 100 participants in the plenary room were enthusiastic about listening to the keynote speakears from Assoc. Prof. Muhammad Imran Qureshi, Ph.D (Universiti Teknikal Malaysia Melaka), Assoc. Prof. Dr. Dahlan Abdullah, ST, M.Kom (Univeritas Malikusalleh).

In closing, we would like to thank the Rector of Universitas Budi Luhur, Universitas Bung Karno, Universiti Teknikal Malaysia Melaka and all parties, for supporting the successfull conference. We would also like to thank all contributors for your good cooperation. Special thanks for OC members for their hard work and patience.

Chair of the Organizing Committee: Dr. Darmawan Napitupulu, ST, M.Kom

International Advisory Board:

1. Prof. Anton Satria Prabuwono, Ph.D (King Abdulaziz University, Arab Saudi)
2. Prof. Dr. Achmad Nizar Hidayanto, S.Kom, M.Kom (Universitas Indonesia, Indonesia)
3. Assoc. Prof. Dr. Richard Napitupulu, ST, MT (Universitas HKBP Nommensen)
4. Dr. Kamalia Kamaruddin (Universiti Teknologi MARA, Malaysia)
5. Dr. Ade Gafar Abdullah, S.Pd., M.Si (Universitas Pendidikan Indonesia, Indonesia)

IOP Conf. Series: Materials Science and Engineering 1071 (2021) 011001 doi:10.1088/1757-899X/1071/1/011001

Member of Editorial Board:

IOP Conf. Series: Materials Science and Engineering 1071 (2021) 011002 doi:10.1088/1757-899X/1071/1/011002

Peer review declaration

All papers published in this volume of *IOP Conference Series: Materials Science and Engineering* have been peer reviewed through processes administered by the Editors. Reviews were conducted by expert referees to the professional and scientific standards expected of a proceedings journal published by IOP Publishing.

- **Type of peer review:** Single-blind(Authors do not know who the reviewers are. The reviewers know who the authors are.

- **Describe criteria used by Reviewers when accepting/declining papers. Was there the opportunity to resubmit articles after revisions?** The members of Reviewer Team reviewed papers by some criterias including technical, quality and presentation criteria such as the relevance to the topics, clear problem proposed, writing structure, paper format requirements, similarity, and english proofread. There is an opportunity given to resubmit the articles after revisions.

- **Conference submission management system:**
  Conference submission management system is use website. Website can be accessed at https://icast.sadaconsultant.ac.id/. Corresponding author process by e-mail ICAST: icast@sadaconsultant.id.

- **Number of submissions received:** 53

- **Number of submissions sent for review:** 53

- **Number of submissions accepted:** 37

- **Acceptance Rate (Number of Submissions Accepted / Number of Submissions Received X 100):** 69.81%

- **Average number of reviews per paper:** 2

- **Total number of reviewers involved:** 6

- **Any additional info on review process:** -

- **Contact person for queries:**
  Dr. Darmawan Napitupulu, ST, M.Kom (darmawan.napitupulu@budiluhur.ac.id)

IOP Conf. Series: Materials Science and Engineering 1071 (2021) 011002 doi:10.1088/1757-899X/1071/1/011002

IOP Conf. Series: Materials Science and Engineering 1071 (2021) 012001 doi:10.1088/1757-899X/1071/1/012001

Exploring Architectural Style of Gianyar and Creating Building Material Innovations

Ida Bagus Gede Wirawibawa[1] and I Dewa Gede Agung Diasana Putra[1]*
[1]Department of Architecture, Faculty of Engineering, Udayana University

E-mail: diasanaputra@unud.ac.id

Abstract. Traditional architectural styles are essential for presenting local identity, so that it is necessary to recognize and explore local architectural identities in an attempt to maintain traditional architectural styles as one of the richness of local culture. Generally, the traditional Balinese architecture, particularly the Gianyar architectural style, is still used by Balinese. The design's application is also attributable to the Balinese local government legislation about using traditional Balinese architectural style in a building. However, the style's design process is an architectural challenge for building actors and people willing to realize that the construction processes usually need more time and costly. Because of this situation, it is considered essential to producing a building material that gives a traditional impression through the efficient use of resources and funds. By introducing a phased approach, this research starts with an analysis of the Gianyar architecture, which is used as a guide in designing traditional-style materials used by the broader communities.

1. Introduction

Globalization has evolved quickly, pushing cities and towns to grow more rapidly. This phenomenon is a global standardization, by sharing and incorporating global information and values between regions, in cultural, social, financial, political and technical aspects [1]. This phenomenon offers regions the difficulty of evolving without neglecting their identity. In the competitive period, losing these identities, including architectural identity, implies losing the identities and values. The difficulty to maintain architectural identity, including the traditional Balinese architecture style, is an architectural challenge for the Balinese in which the construction of traditional Balinese architecture is complicated and costly. Therefore, it is essential to create a new construction model, including the materials, to easily apply the traditional Balinese architectural style in their new buildings.

Learning history is a means to scrutinize identity when it makes a relationship between what has allegedly been in the past and what is currently happening. The construction of history is a meaningful universe of events and stories for a person or a collection of subjects [2]. Moreover, as the object of this creation process stems from a topic living in a particular social environment, we might say that history is an example of the present in the past that is used as a region's identity.

Architectural practice is one of the significant components that express the distinctive identity of a society in a particular region. Four elements distinguish one place or area from others, including spatial qualities (including architecture) and culture or history [3]. As a significant kingdom in Bali in the 15-19th century, Gianyar was not only the centre of government but also the centre of cultural and ritual activities. The vital role of Gianyar in the past produced various cultural practices, including architecture [4]. Cultural and ritual practices, including architecture, were ways to express the strength of a kingdom's influence and to express its distinctive identity [5], [6].

The localization and diversity of Balinese architectural dialects have been influenced by Bali's political agenda, primarily since the 18th century when the Majapahit Kingdom from Java conquered Bali, in which the palace that is the central government of the kingdom is called *puri*. Bali was then ruled from the single primary *puri* (palace) of Klungkung, and the other *puri*s such as Badung, Bangli, Tabanan, Karangasem and Gianyar were described as secondary independent palaces [5]. To express their superiority, the primary and secondary kingdoms performed state cultural practices in their territories to reveal their status differences [5], [6]. Every palace, including Gianyar, developed independently the Balinese culture that led to the differences in the implementations of cultural practices, including the architecture. They expressed themselves and created boundaries with others. In this context, tradition and culture can be seen as a resource to develop boundaries and identity [7]–[10].

The application for the architecture is also attributed to the laws of the local authority in Bali. The design phase of the style, however, is an architectural obstacle for construction actors and individuals who have a desire to consider how it is restricted by time and expense. This condition makes it necessary to produce a building material through the efficient use of resources and funds but still expresses the traditional style. This study begins by examining Gianyar architecture as a reference for the conception of traditional models that have been used by communities and followed by the production of material designs for the traditional style of buildings.

2. The Architectural Style of Gianyar

The Kingdom of Gianyar was seeking to put its territories into a new image as a binding and creative identification, with the different problems and influences of other kingdoms in Bali. The Gianyar Authority sought to forget the boundaries of previous forces, driven by complex cultural and political motivations [11]. Around the same time, a contemporary architectural production has demonstrated how this new area has created something massive and monumental. In that case, technology has been used to portray general concepts about race and the great society.

Three sections are the shape of the *kori agung* in Bali, including the *kori agung* of the royal house in Gianyar. The head is on the top of the structure, while the middle part is the body, and the lower part is the legs. There are three layers of the head and a small *bentala* or *mudra* on the top; one main structure and one or two *lelengen*s on their legs. The body is equipped carved wood doors, which is the way for the offerings to go in and out during ceremonies [12]. The legs on which *kori agung* are fitted with brick steps (Figure 1). This *kori agung* was an architectural production to establish a local identity for the royal family [7]–[9]. This identity shows a generational disparity between members and non-members [10]. In this tradition, the royal houses are seen as the crossroads of the history of goods and customs. This intersection is an occurrence between many vernacular buildings [13].

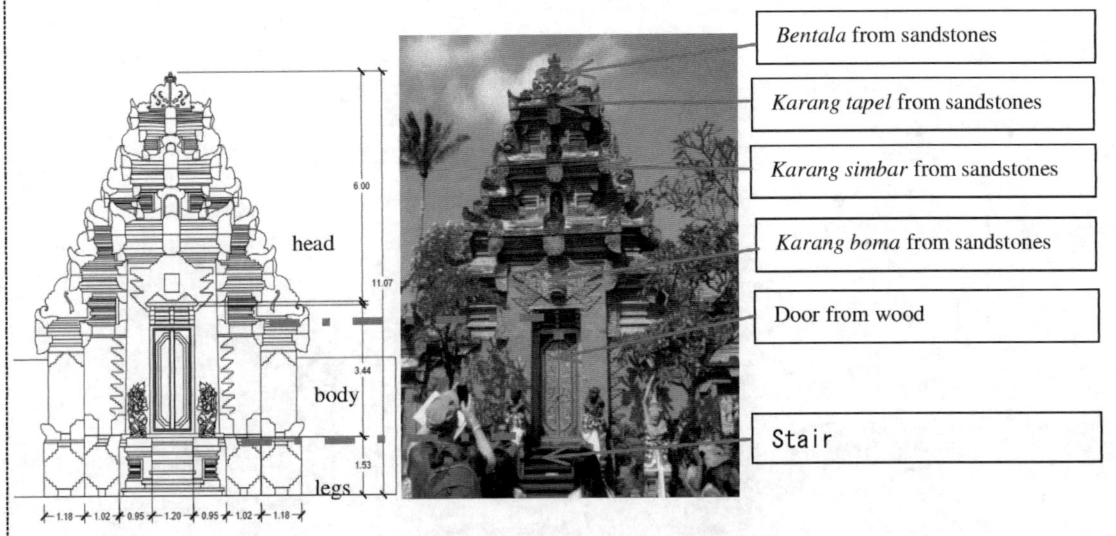

Figure 1.A *kori agung* of a royal family house in Gianyar

As a royal family house, the gates are made of a combination of red brick and sandstone. In general, bricks are the main structure, while the ornament is generally made from sandstone [14]. The sandstones are attached to brick and carved in detail with plant and animal motifs as ornaments. Ornament is a Latin word, *onare*, meaning decoration. The concept initially encompasses elements and decorations adapted from natural leaves and created from them [15]–[18].

3. The Model of a contemporary building material adopting traditional architectural style.

The existing bricks have a similar form, so that if the people want to make the traditional style, they should call traditional builder experts. The limitation of the brick's forms has created difficulty for Balinese communities, including in Gianyar, to applied traditional style in the contemporary buildings. Therefore, it is necessary to make various forms to fulfil the use of decoration as required by traditional buildings. In previous research about brick materials, *silicium* is used to manufacture the *calcium silicate hydrate* that acts as an adhesive [19]. A further analysis was made on the topic of researches [20], [21], containing bagasse-ash or equivalent additive content. To avoid combustion, sand and water were used in the moulding process on the hydraulic machinery to produce interlocking bricks [22], [23]. Many researches have been carried out on the invention of interlocking brick with a bagasse ash mixture using clay as the base of a material mixed with cement [21].

The interlocking brick is produced using a moulding machine that requires unskilled labourers. This brick is so dense that the walls look very sturdy and not too thick to build the traditional Balinese walls. By showing the positive and negative elements of the upper, lower and sides of brick surfaces, the bricks can be dried and can automatically be levelled when the joints are finished. This interlock system has been designed to make easier construction process for unskilled labourers [24]. A pyramid locking mechanism with a hole in the centre has the benefit of the interlocking brick style. This method was deemed better because the transition of the excitement of force can be uniformly spread horizontally [25].

This application needs to be done to simplify and accelerate the construction that still impresses these traditional buildings. To get the desired form, three basic forms must be made. The first form is the body form. This form has the same shape as a brick in general, and it is just that the size of the module is bigger so that it can speed up the installation process but can still be lifted by a builder. The second form is the legs. This form will later function as the legs of the wall, which has width greater than the body and has little ornamentation. This leg form can be made the same as the wall head form. The third form is the pole form. This form functions as a corner meeting of the walls with a certain thickness (Figure 2). Here there are negative and positive forms (opposites). Besides that, the legs of the pillar and the head are also needed.

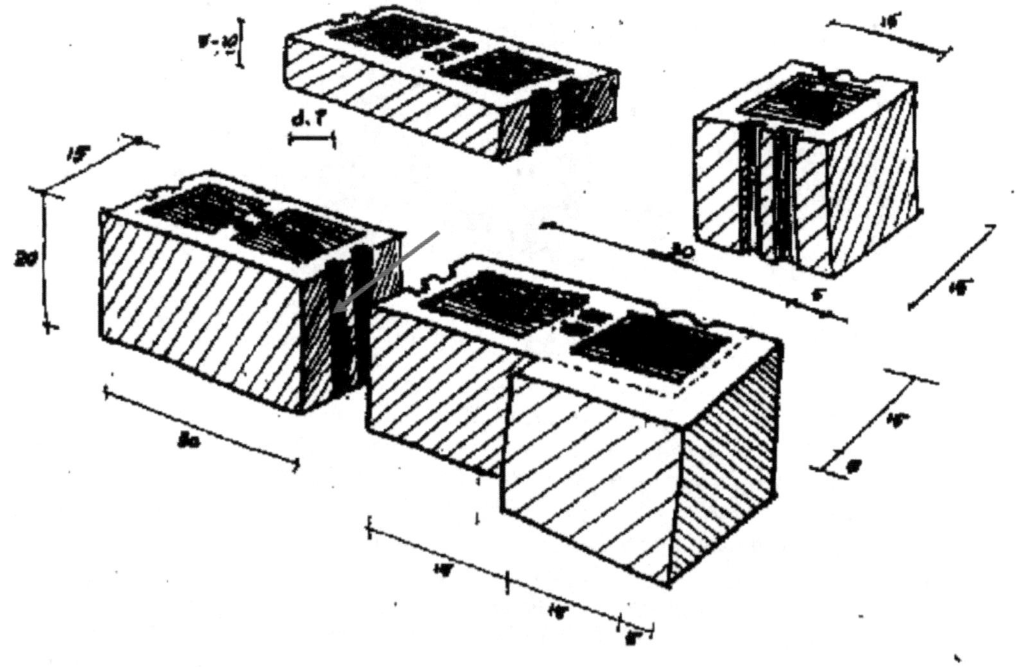

Figure 2. The alternative form of the bricks

The existence of different forms will undoubtedly affect the mould that is made. In this case, three moulds should be required, which have different sizes from each other. This condition will certainly make it quite expensive to make these different prints. Besides that, the mould also requires a place to work on it. In this case, a wider area is needed for the work areas. In this situation, purchasing material would minimize the expense, which, for the construction of the projects, would guarantee compliance with the standard of the already developed material and material of the sort on the market. Several experiments used waste materials to mitigate the cracks in the project system [26], [27]. The bricks industry is the most highly-informed industry for technical waste absorption by the large volume of raw material used in manufacturing and the large volume of finished goods [28].

To reduce costs and use the land for making the mould, it is necessary to modify the mould to be flexible and make printing for various forms. The most likely thing to do is to use a mould where some of the components can be removed. It is necessary to think more deeply in deciding the form of this print, which on the one hand should facilitate the work, especially in production, but also reduce costs and saving time. To get the three things above together is certainly not possible. Therefore what needs to be done is to create the most optimal form so that the production results are maximized.

Seeing the shape requirements of the bricks, two moulds are needed, both of which must be flexible. These shapes correspond to their functions. The first is the shape of walls, and the second is the shape of the pillars. These two forms of moulds must produce a brick that will fit perfectly into each other. Furthermore, each will produce two types of forms, namely the shape of the legs, including the head and the other the body shape.

4. Conclusions

In general, brick technology can be modified to produce bricks that have a traditional feel through amalgamation in ways commonly practiced by construction practitioners in Bali so far. This traditional nuance can be developed by investigating the traditional architecture of the *kori agung* of the royal houses in Gianyar. The development of bricks with traditional nuances is made to facilitate the work of traditional buildings and does not require special skills for the installers. This traditional Balinese architectural style brick can produce a traditional Balinese impression without producing much waste when working.

References

[1] HKU, 2012, Liberal Studies Teaching Kit for Senior Secondary Curriculum: Globalization and Localization of Architecture and Urban Planning. Hong Kong: Faculty of Architecture, the University of Hong Kong., Hongkong.
[2] Friedman J, 1992, The Past in the Future: History and the Politics of Identity *Am. Anthropol.* **94**, 4 p. 837–859.
[3] Amundsen A B, 2001, Articulations of identity: A methodological essay and a report on Askim and Tidaholm *NoordXXI Rep. no. 19. Retrieved from – http -//www. Noordxxi.nl/.*
[4] Agung I A A G, 1991, *Bali in the 19th Century* Yayasan Obor Indonesia.
[5] Geertz C, 1980 *Negara* Princeton University Press.
[6] Nordholt H G C S, 1986, Bali: Colonial conceptions and political change 1700-1940. From shifting hierarchies to'fixed'order.
[7] Derek and Japha V, 1991, Identity through detail: an architecture and cultural aspiration in Montagu, South Africa, 1850-1915 *TDSR, II* p. 17–33.
[8] Hall S, 1990, Cultural identity and diaspora, in *Identity, community, culture difference*, J. Rutherford, Ed. (London: na), p. 222–237.
[9] Proshansky H M, Fabian A K and Kaminoff R, 1983 Place-identity: Physical world socialization of the self *J. Environ. Psychol.*
[10] Brubaker R and Cooper F, 2000, Beyond" identity" *Theory Soc.* **29**, 1 p. 1–47.
[11] Gandhi L, 2019, *Postcolonial theory: A critical introduction* Columbia University Press.
[12] Putra I D G A D and Wirawibawa I B G, 2020, The Balinese Palaces in Gianyar : Representing Authority Power and Creating Territorial Identity **14**, 8 p. 132–150.
[13] Lozanovska M, 2011, Holy days after migration in *Proceedings of the 2nd International Conference on Intangible Heritage: Sharing Cultures* p. 459–469.
[14] Dwijendra N K A and Suyoga I P G, 2020, Analysis of symbolic violence practices in balinese vernacular architecture: A case study in Bali, Indonesia *Int. J. Innov. Creat. Chang.* **13**, 5 p. 184–194.
[15] Meyer F S, 1974, *A handbook of ornament* Рипол Классик.
[16] Sudara I G N, 1983, Ornamen Bali (Kumpulan Ragam Hias). Penerbit: Sekolah menengah Seni Rupa Denpasar: Denpasar Bali.
[17] Gelebet I N, 1998, *Arsitektur Tradisional Bali (Balinese Traditional Architecture)* Denpasar: Bappeda Bali.
[18] Satria M W and Putra I D G A D, 2020, The Kori Agung Character of Heritage Temples: The Architectural References of Klungkung Identity *J. Soc. Polit. Sci.* **3**, 1.
[19] Sudarsana I K, Budiwati I A M and Wijaya Y A, 2011, Karakteristik Batu Bata Tanpa Pembakaran Terbuat Dari Abu Sekam Padi Dan Serbuk Batu Tabas *J. Ilm. Tek. Sipil* **15**, 1 p. 93–101.
[20] Abdurrohmansyah Adha I and Ali H, 2015, Studi Kuat Tekan Batu Bata Menggunakan Bahan Additive (Abu Sekam Padi, Abu Ampas Tebu & Fly Ash) Berdasarkan Spesifikasi Standar Nasional Indonesia (SNI) *Jrsdd* **3**, 3 p. 2303–2314.
[21] Alamsyah A A *et al.*, 2017, Pemanfaatan Abu Ampas Tebu (Bagasse Ash of Sugarcane) Sebagai Bahan Pengisi (Filler) Dengan Variasitumbukanpada Campuran Aspal Panas Atb

(Asphalttreatd Base) *Semin. Nas. Teknol. dan Rekayasa* p. 1–10.
[22] Habsya C, Rahmawati A and Sumarni S, 2004, Lockbrick Modular Beton Untuk Alternatif Bahan Dinding *Sinektika* **14**, 2 p. 234–242.
[23] Kafrain I G Y, 2018, Dinding Bata Interlock Pulutan *J. Ilm. Realt.* **14**, 1 p. 75–80.
[24] Putra I D G A D, 2003, Teknologi bahan bangunan rumah murah bergaya tradisional bali *J. Permukim. Natah* **1**, 1 p. 44–51.
[25] Erwanto Z, Pranowo D D, Holik A, Amin M S and Darmawan F, 2020, The Innovation of Interlock Bricks with A Mixture of Bagasse Ash Without Combustion *IOP Conf. Ser. Mater. Sci. Eng.* **854**, 1 p. 1–11.
[26] Cengizler H, Çiçek T and Tanrıverdi M, 2012, A Brief Overview of Fly Ash Brick Production *Proceeding XIII. Int. Miner. Process. Symp. 10 – 12 Oct. 2012, Bodrum, Turkey* p. 983–992.
[27] Dinesh W G and Laxmikant N V, 2014, Strength Characteristics of Different Types of Bricks. *Int. J. Sci. Res. (IJSR). Oct 2014; 3(10)* p. 2247–2252.
[28] Andreola F, Barbieri L, Lancellotti I and Pozzi P, 2005, Recycling industrial waste in brick manufacture. Part 1. Journal of Material Construction, 1: 5-16. p. 5–16.

IOP Conf. Series: Materials Science and Engineering 1071 (2021) 012002 doi:10.1088/1757-899X/1071/1/012002

Evaluation of Internal Forces and Support Reaction of Column and Shearwall in 15th Building Design

D R Wiyono[1], R Milyardi[1*] and Y A Pranata[1]
[1]Department of Civil Engineering, Faculty of Engineering, Universitas Kristen Maranatha, Bandung, West Java, Indonesia

E-mail: roi.milyardi@maranatha.edu

Abstract. In the case of 15th building, the structure type to earthquake resistant code is dual system reinforced concrete consists of frame and shear wall. The structure is analysis by dynamic analysis for whole structure and compare base shear by static analysis. By several trial position of shear wall can give first and second mode is translation and the third mode is rotation. In this case planar shear wall is used and placed in two direction of axis. The category of structural system is Other Structure for calculation of fundamental period (T). Fundamental Period (T) of structure is more than T maximum, so value of static seismic coefficient (C_s) taken from T maximum. Each of important element is column where the dimension and reinforcement are must satisfied for requirements. Shear wall have joint together with others shear wall, so one node used by several shear wall, make duplication forces in support reaction. For foundation design we can used 1st story internal forces in finding forces and compare them with after correction of support reaction caused duplication node in used with several shear walls. This exercise could solve the problem to anticipate duplication forces in several planar shear wall using one foundation assembly from several shear walls. The results of preliminary design column based on axial forces is 23,12 % lower than internal forces. From preliminary design based on axial forces. The dimension of preliminary design columns are lower than ideal dimension with 3 % reinforcement ratio. The result of difference between internal forces columns and support reaction columns is 12,99 %. The result of difference between internal forces shear walls and support reaction shear wall is 0,80 %.

1. Introduction

In the case of 15th building the structure is dual system reinforced concrete consist of frame and shear wall. The structure is analysis by dynamic analysis for whole structure and compare base shear by static analysis. By trial position of shear wall can give first and second mode is translation and the third mode is rotation. In this case planar shear wall is used and placed in two direction of axis. The category of structural system is Other Structure for calculation of fundamental period. Period of structure is more than T maximum, so value of Cs taken from T maximum. Dimension and reinforcement of column is very important because must be resist a large portion of forces and have performance as strong column weak beam. Evaluation of column by compare preliminary design column and internal forces because there are several aspects in structure such earthquake load can change dimension to large from referred by dimension from preliminary design. The effect of axial load and moment will be evaluation for to have prediction about reinforcement or dimension, where ideal reinforcement column is about 3 %, if do not want use mechanical connection to joint reinforcement bar [1], [2]. Using shear wall can reduce building period and story drift because behavior of fixed joint at restraint. Evaluation of shear wall by compare internal forces especially axial forces in shear walls with support reaction. In support reaction gives forces in every node, this will be duplication forces in same node for two or three or four walls connected to this node. For foundation design because there are several walls so each wall will give forces to one pile cap assembly to several walls. Correction forces in support reaction can referred to internal forces of shear walls, and by modelling separate foundation with forces from several shear walls will take number of piles.

2. Literature Study

In high rise buildings, combination of shear walls and frames in normally provide the required stiffness and strength to withstand lateral loads. Shear walls normally are much stiffer than the frames system to take of lateral load more than frame system [3], [4].Because of larger stiffness, the contribution of the frame system in resisting lateral load is usually ignored. This practical procedure must be corrected because there is important that the effect of the frames be considered in seismic resistant building [5].

A rigid frame consist of vertical columns and horizontal beams, bends predominantly in a shear mode shown in Figure 1 (a) [6] and a shear wall in a bending mode, i.e, as a cantilever, as illustrated in Figure 1(b). In some building structure (stair opening, elevator shafts) which reinforced concrete walls normally show the behavior [7]. The combination between rigid frame and shear wall interacting will tend to deflect in a bending mode [8]. The analysis is comparatively simple when all vertical units of a structure show the same behavior under lateral load which the seismic system in all rigid frames or all shear walls. The load can be distributed to the units proportionally to their stiffnesses. The difference in behavior under lateral load, in combination with the in-plane rigidity of the floor slabs, causes nonuniform interacting forces to develop when walls and frames are present Figure 1 (c). This makes the analysis more complex.

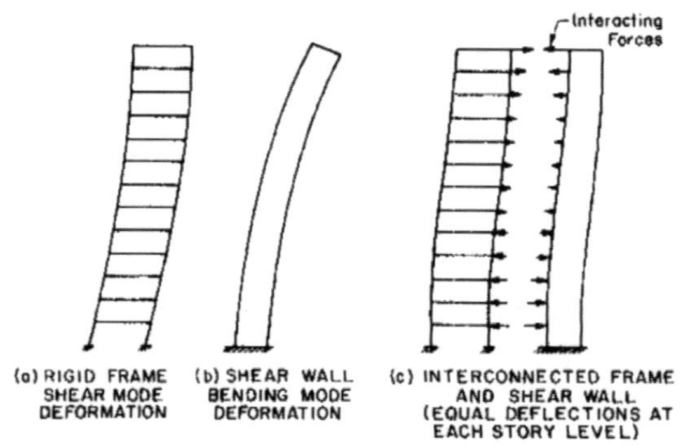

Figure 1. Deformation modes [6]

For analysis, the floor slabs are normally considered to be fully rigid within their own planes. This means that there will be no relative movement between the vertical units at each story level.

3. Analytical Model

In this study, building model is 15 story dual system with other Structure System. The function of this building is hotel in Jambi City with seismic parameter Ss = 0,7 g, S1 = 0,4 g with site class D [9]. Material property of the building of concrete is f'_c = 28 MPa and steel reinforcement is f_y = 420 MPa for plate, beam, column and wall. The gravity load for this building functioned as hotel facilities [10]. The structure is given in the Figure 2, and in Figure 3a shows point label and line label to indicate the support reaction output location in building[11].

IOP Conf. Series: Materials Science and Engineering 1071 (2021) 012002 doi:10.1088/1757-899X/1071/1/012002

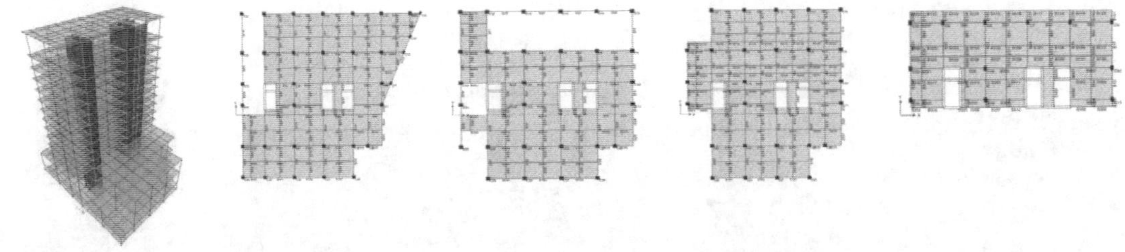

(2a) 3D Buildig Model Floor Plan | (2b) 1st Floor Plan | (2c). 2nd Floor Plan | (2d). 3rd Floor Plan | (2e). 4th -15th Typical

Figure 2. Building Model and Floor Plane Data

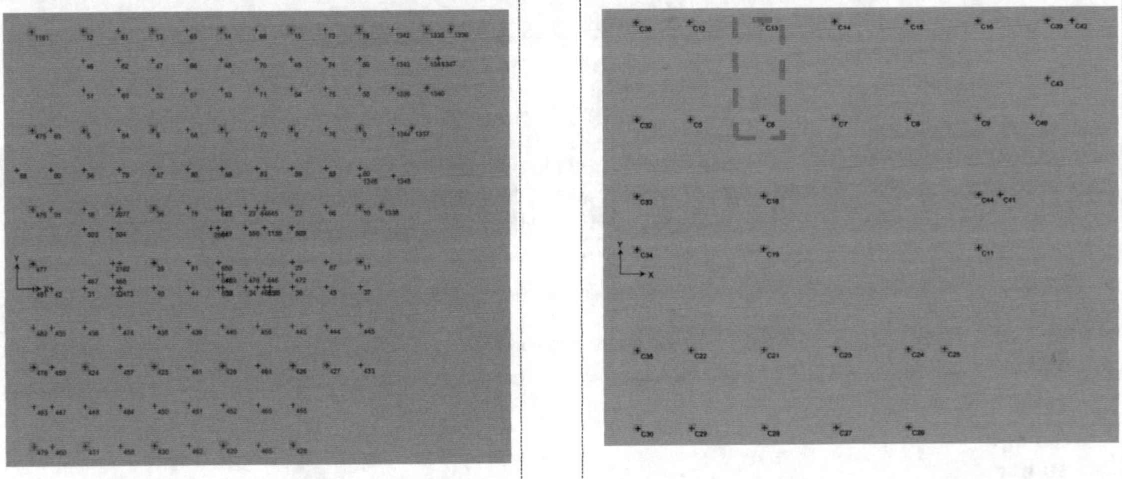

(3a) Point label of support reaction | (3b) Line label of column

Figure 3. Building Point label ID Data

4. Result and Discussion

Output ETABS and preliminary dimension of column is shown in table below, it is known that axial force ETABS bigger 23,12 % than axial force preliminary show in 0.

Table 1. Comparation of Colum Axial Force.

Story	Column	P Etabs (kg)	P Prelim (kg)	Magnify Factor	% Difference
STORY1	C6	1212399	999591.6	1.21	21.29
STORY1	C18	1019156	815688	1.25	24.94
		Average			23.12

In 0 and 0 shown column dimension with variance reinforcement for getting ideal dimension.

Table 2. ETABS Column dimension with variance reinforcement.

Column ID	Reinforcement Rebar Area, As (mm²)	Column Section Area											
		Parameter	**1.00 % As**	**1.50 % As**	**2.00 % As**	**2.50 % As**	**3.00 % As**	**3.50 % As**	**4.00 % As**	**4.50 % As**	**5.00 % As**	**5.50 % As**	**6.00% As**
C6	294.38	Section Area, Ag (mm²)	29438	19625	14719	11775	9813	8411	7360	6542	5888	5352	4906
		Dimension (cm)	172x 172	140x 140	121 x 121	109 x 109	99 x 99	92 x 92	86 x 86	81 x 81	77 x 77	73 x 73	70x 70
C18	169.76	Section Area, Ag (mm²)	16976	11317	8488	6790	5659	4850	4244	3772	3395	3087	2829
		Dimension b x h (cm)	130x130	106x106	92x 92	82x 82	75 x75	70 x70	65 x65	61 x61	58 x58	56 x56	53x53

Table 3. Preliminary Column dimension with variance reinforcement

Column ID	Reinforcement Rebar Area, As (mm²)	Column Section Area											
		Parameter	**1.00 % As**	**1.50 % As**	**2.00 % As**	**2.50 % As**	**3.00 % As**	**3.50 % As**	**4.00 % As**	**4.50 % As**	**5.00 % As**	**5.50 % As**	**6.00% As**
C6	294.38	Section Area, Ag (mm²)	4841	4617	4413	4227	4055	3897	3751	3615	3489	3371	3261
		Dimension (cm)	70x 70	68x 68	66x 66	65x 65	64 x 64	62 x 62	61 x 61	60 x 60	59 x 59	58 x 58	57x 57
C18	169.76	Section Area, Ag (mm²)	3950	3768	3601	3449	3309	3180	3061	2950	2847	2751	2661
		Dimension b x h (cm)	63x 63	61x 61	60x 60	59x 59	58 x 58	56 x 56	55 x 55	54 x 54	53 x 53	52 x 52	52x 52

From the above table can get the magnify reinforcement between preliminary and output ETABS as shown in Figure 4 below. The ideal dimension for column C18 and column C6 is shown in Table 4 below.

IOP Conf. Series: Materials Science and Engineering 1071 (2021) 012002 doi:10.1088/1757-899X/1071/1/012002

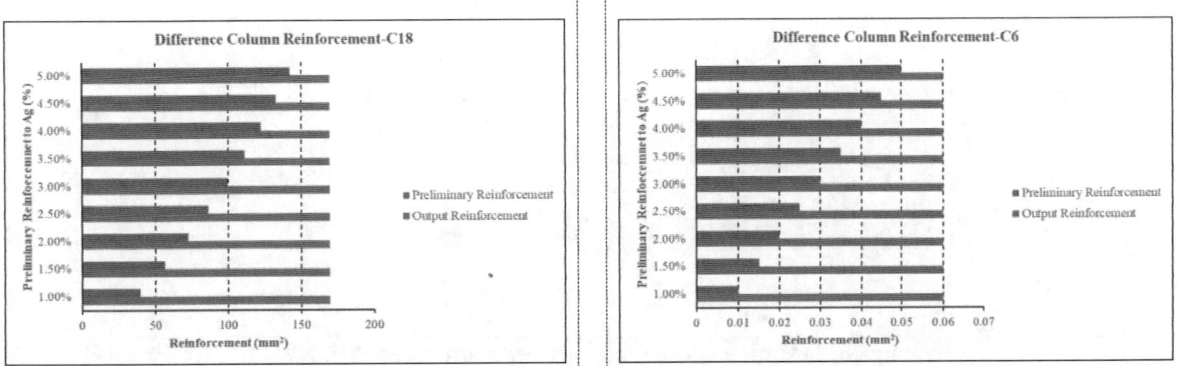

Figure 4. Different Reinforcement Column C18 and C6

Table 4. Ideal dimension column C18 and C6

Ideal Dimension C18			Ideal Dimension C6		
Ag (cm^2)	As (cm^2)	Magnifying Factor	Ag (cm^2)	As (cm^2)	Magnifying Factor
5658	0.03	1.43	9812	0.03	2.03
75x75			100x100		

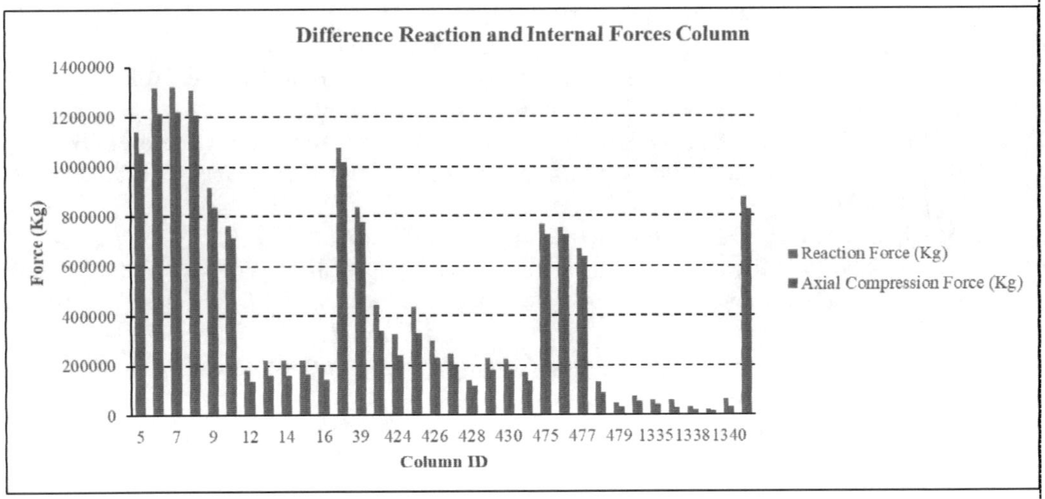

Figure 5. Difference Reaction and Internal Forces Column

Figure 5 shown that ideal dimension is 43,2 % in C18 and 202, 7 % in C6 upper than preliminary dimension.

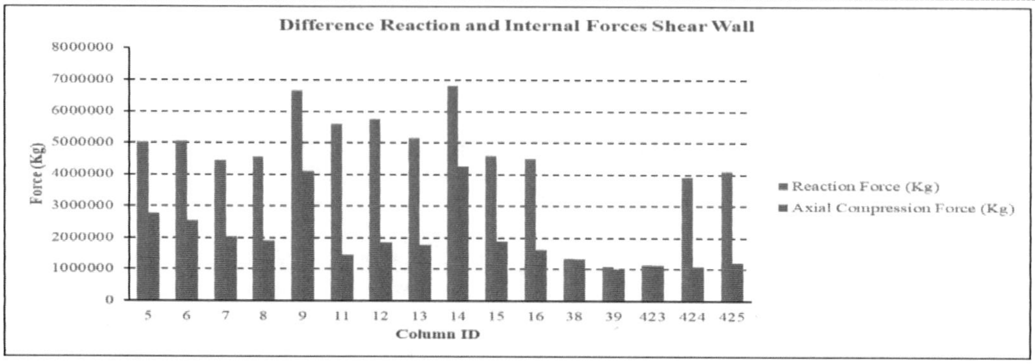

Figure 6. Difference Reaction and Internal Forces Shear Wall

Figure 6 shows that difference column reactions are 12,99 % bigger than column internal forces, and that difference shear walls are 0,89 % then shear walls internal forces.

5. Conclusion

The result of preliminary design based on axial forces is 23,12 percent lower than internal axial force. The dimension of preliminary design C18 column is 43,2 % lower than reality dimension and C6 column is 202, 7% lower than reality dimension with 3 percent reinforcement ratio. The result of difference between internal forces columns and support reaction columns is 12,99 %. The result of difference between internal forces shear walls and support reaction shear walls is 0,89 %.

References

[1] J. G. Wight, J. K. ; MacGregor, *Reinforced Concrete: Mechanics and Design, 6th Edition*. New Jersey: Pearson Education, Inc, 2012.

[2] American Concrete Institute, *2011 Building Code Requirements for Reinforced Concrete (ACI 318-11)*. American Concrete Institute, 2011.

[3] L. Budiono, B; Supriatna, *Studi Komparasi Desain Bangunan Tahan Gempa dengan Menggunakan SNI 03-1726-2002 dan RSNI 03-1726-201X*. Bandung: Penerbit ITB, 2011.

[4] B. et al Budiono, *Contoh desain bangunan tahan gempa dengan sistem rangka pemikul momen khusus dan sistem dinding struktur khusus di Jakarta*. Bandung: Penerbit ITB.

[5] I. H. Imran, *Perencanaan Struktur Gedung Beton Bertulang Tahan Gempa*. Bandung: Penerbit ITB, 2010.

[6] A. MacLeod, "Shear Wall-Frame Interaction A DESIGN AID," 1970.

[7] Wiyono, Daud R.; Milyardi, Roi; Lesmana, "The Effect of Shear Wall Configuration on Seismic Performance in the Hotel Building," *2nd Int. Jt. Conf. Adv. Eng. Technol. (IJCAET 2017) Int. Symp. Adv. Mech. Power Eng. (ISAMPE 2017*, 2018.

[8] P. Somers, *Reinforced Concrete-Instructional Materials Complementing FEMA P-751, Design Examples*. FEMA, 2012.

[9] Badan Standarisasi Nasional, *Tata cara perencanaan ketahanan gempa untuk struktur bangunan gedung dan non gedung (SNI 1726:2012)*. Jakarta: Badan Standarisasi Nasional (BSN), 2012.

[10] American Society of Civil Engineering, *Minimum Design Loads and Associated Criteria for Buildings and Other Structures (ASCE 7-10)*. American Society of Civil Engineering, 2010.

[11] Computers and Structures Inc., *CSI ETABS, Concrete Shearwall Design Manual*. University Avenue.

IOP Conf. Series: Materials Science and Engineering 1071 (2021) 012003 doi:10.1088/1757-899X/1071/1/012003

Application of Anthropometry Methods in Ergonomic Chair Redesign to Prevent Fatigue A Case Study UKM Lestari Jaya, Tulungagung

Silviana[1], A Hardianto[1]*, D Hermawan[2] and Abdurahman[3]
[1]Lecturer, Industrial Engineering Dept, University of Widyagama Malang, Indonesia
[2]Lecturer, Mechanical Engineering Dept, University of Widyagama Malang, Indonesia
[3]Student, Industrial Engineering Dept, University of Widyagama Malang, Indonesia

E-mail: andyhardian@gmail.com

Abstract. Small and Medium Enterprises (UKM) Lestari Jaya is one of the home industries in Pucuk Village, Tulungagung Regency that produces chips from processed cassava ingredients which are commonly called samiler. These samiler chips are produced using manual methods so that the production process still depends on workers. So far, the work chairs used for military production, especially in the printing process, are not ergonomic and cause fatigue. This work chair is said to be not ergonomic in terms of the form of the backrest that is far from the back of the worker, the size that does not fit the worker's body shape, the chair material is made of plastic and the resistance of the chair in accepting body loads. Based on data from the Nordic Body Map questionnaire distributed to workers, chairs that are not ergonomic can cause fatigue in workers. Based on the above problems, a study was carried out to redesign an ergonomic work chair with the aim of reducing the risk of fatigue in workers. The anthropometric calculation of percentiles is used to determine chair size. The data is in the form of the 95th percentile for the seat height (45 cm) so that it can be used for people who have long or short legs, the 95th percentile for the width of the seat base (44.02 cm) so that the size of the seat base is wide, the 50th percentile is for the back height (56.58 cm) so that it can be used for tall and short people, the 50th percentile is for the length of the seat base (44.93 cm) so that people with large and small popliteal bottoms can use this chair, and the 50th percentile is for the height of the armrests (23, 95 cm) to fit the average human hand size. The advantages of this chair design are that it is stronger, the size of the chair is in accordance with the worker's body size and the addition of functions, namely that the chair's height can be raised or lowered according to the worker's body size

1. Introduction

Ergonomics is the science, art and application of technology to harmonize or balance all facilities used for activities and breaks with human capabilities and limitations both physically and mentally so that the overall quality of life becomes better (Tarwaka, 2004). Ergonomics is closely related to anthropometry in harmonizing or balancing all the facilities used in work. Because anthropometry is a natural measure of the human body both in carrying out static activities (actual size) and dynamic (adjusted to work) (Wignjosoebroto, 2003). By knowing the dimensions of the worker's body through anthopometric calculations, a design of work equipment, work stations and products can be made in accordance with the dimensions of the worker's body so that it can create work comfort, health and safety. In an industry or business, workers' comfort must be considered related to the facilities used, workload that does not cause fatigue, and comfort when doing work. In Tulungagung to be precise, has one of the Lestari Jaya Small and Medium Enterprises (UKM) which is engaged in the snack industry, namely military production. Samiler is a cracker made from processed cassava-based ingredients that are mashed, printed, dried and fried. One of the production activities is carried out by workers sitting in a work chair, namely when making dough. So far, using chairs that are not ergonomic because they do not match the worker's body size can cause fatigue after work. Apart from not providing comfort and causing pain, the work chairs used at UKM Lestari Jaya are made of plastic material which if the body load is too heavy it will cause the lower chair support to curve so that in the long term it can also cause work accidents. The following is a picture of a sitting position in a work chair that is not ergonomic.

Figure 1. Samiler Pressing process
Source UKM Lestari Jaya

Figure 1 represent the press process carried out in a non-ergonomic chair position. An improperly designed chair leads to a sitting position which increases muscle tension and balance control leading to fatigue and discomfort. Increased muscle pressure and disturbed balance control can have an impact on worker fatigue.

2. Literature Review

Industrial ergonomics is the application of various sciences that link human performance with a work system consisting of workers, jobs, equipment and equipment, workplaces and their environment, Definition of Anthropometry Anthropometry comes from the words anthropos and metricos. Anthropos means human and metricos means measure. Anthropometry is the measurement of the human body naturally both in carrying out static activities (actual size) and dynamic (adjusted to work). Anthropometry is a science that deals with the measurement of dimensions and other characteristics of the human body such as volume, center of gravity and mass of human body segments[1]. The size of the human body varies widely, depending on age, sex, race, occupation and period from time to time. Measuring the dimensions of the human body is the most important part of anthropometry because it will be the basic data for preparing the design of various equipment, machines, processes and workplaces. We often find several industries in Indonesia that the design of work equipment and work stations is not in accordance with the dimensions of the worker's body[2]. For example, industrial display designs that are difficult for operators to read. Another example is the operator doing work where the workpiece is on the floor, so the operator performs activities with a bent work attitude or sitting in a short chair[3]. This work attitude will cause premature fatigue and even injury to the spine and cervical spine. Another case we often encounter is the design of classroom equipment for Elementary Schools (SD), where the design of tables and chairs for grades one to six has the same size. This condition makes students in grades one to grade three who have body dimensions smaller than grade four to grade six students will feel uncomfortable in following lessons. Several studies related to the design of elementary school desks and chairs have been carried out. The research carried out is in the form of proposed designs and implementation of ergonomics-based designs to reduce musculoskeletal complaints However, the results of research This has not received a positive response from stakeholders, so we often encounter table and chair designs that are not in accordance with the dimensions of the elementary school student body[4] , [5]. Today the application of anthropometric data does not only concern the characteristics of the equipment, equipment and everything used in carrying out work activities, but also the design of work stations. In order for the design of work stations to be comfortable to use for activities, it is necessary to carefully consider including the use of anthropometric data[6]. The design of a work station requires knowledge of the range limits of the human body, known as the normal working area and the maximum working area.

2.1. Anthropometric Calculations
There are several anthropometric calculations that need to be done in data processing

2.2. Data Normality
The data normality test aims to determine whether the collected data follows a normal distribution. Normality testing will direct advanced statistical techniques to be used for decision-making tests, Data normality tests were carried out using the Shapiro-Wilk analytical test method. The Shapiro-Wilk test

$$\bar{X} = \frac{\sum Xi}{n} \tag{1}$$

method is used when the sample size is small (≤50)[7], [8]

$$N' = \left[\frac{\frac{K}{S}\sqrt{N(\sum Xj^2) - (\sum Xj)^2}}{\sum Xj}\right] \tag{2}$$

One of the factors that affect occupational health is related to ergonomics or work attitudes such as repetitive work and non-ergonomic positions. Apart from that the hours of activity are Nordic Body Map, One of the tools used to determine the description of Musculoskeletal Disorders is the Nordic Body Map questionnaire. Nordic Body Map is a questionnaire in the form of a body map containing data on the body parts complained by workers. The Nordic Body Map questionnaire is the most frequently used questionnaire to find out workers' discomfort, and this questionnaire is most often used because it is standardized and neatly structured[9].

3. Results and Discussion

3.1. Data Collection
The data collection stage is the stage where the researcher will look for the data needed to be processed. The data collected by researchers are data on the production process identification of the work chair used and the sitting position of the worker using the old work chair, At this stage the researcher will process the data that has been taken at the data collection stage. The data obtained will be a measure used in designing a work chair through anthropometric data processing. The data normality test is used to determine whether the data in a group or variable based on the test is normal or not. Data is said to be normal with a failure rate of 5% or 0.05 using SPSS software[10], [11]. That is, if the data with a significance ≥ 0.05 is normal data, while if the significance is ≤ 0.05 then the data is considered abnormal and must take data again. There are two types of normality tests, namely the first Kolmogorov for large samples ≥ 50 data[12], [13], [14]. Then the second is Shapiro for small samples ≤ 50 data. In this study, because the data were below 50, the Shapiro normality test was used. The calculation results can be seen in table 1.

IOP Conf. Series: Materials Science and Engineering 1071 (2021) 012003 doi:10.1088/1757-899X/1071/1/012003

Table 1. Results of the data normality test calculation

Tests of Normality

	Kolmogorov-Smirnov[a]			Shapiro-Wilk		
	Statistic	**Df**	**Sig.**	**Statistic**	**df**	**Sig.**
Tpo	.137	40	.056	.960	40	.163
LP	.126	40	.111	.950	40	.079
TSP	.112	40	.200*	.945	40	.051
Ppo	.108	40	.200*	.969	40	.333
TSD	.107	40	.200*	.970	40	.350

*. This is a lower bound of the true significance.
a. Lilliefors Significance Correction

The next test is to find out whether the data obtained is in a controlled state or not. Data that is within the specified control limits, namely Upper Control Limit and Lower Control Limit can be said to be in a controlled state, on the other hand, if the data is outside the Upper Control Limit and Lower Control Limit, then the data is said to be uncontrolled. Data that is in an uncontrolled state will be discarded and then tested again for uniformity until there are no more data that are outside the Upper Control Limit and Lower Control Limit

Table 2. Results

No	Description	SD (δ)	BA	BB
1	Tpo	3.986	45.148	29.202
2	LP	2.957	44.139	32.311
3	TSP	2.908	63.391	50.759
4	PPo	3.846	52.616	37.234
5	TSD	2.688	29.237	18.573

3.2. Percentile

The percentile data is used to determine the size of the chair that will be made according to anthropometric data calculations. Percentile shows the simulated average person data in percentage form. For example, the 95-th percentile will indicate 95% of the population will be at or below that measure; whereas the 5-th percentile will show 5% of the population will be at or below that measure. In anthropometry, the 95-year number represents the "largest" human size and the 5-th percentile represents the "smallest" size[15].

TPo

Percentil 5 $= X - 1{,}645\delta$

$= 37{,}175 - 1{,}645\ (3{,}986435)$

$= 30{,}61731$

Percentil 50 $= 37{,}175$

Percentil 95 $= X - 1{,}96\delta$

$= 37{,}175 + 1{,}96\ (3{,}986435)$

$= 44{,}98841$

LP

Percentil 5 $= X - 1{,}645\delta$

= 38,225 – 1,645 (2,956848)
= 33,36099

Percentil 50 = 38,225

Percentil 95 = X – 1,96δ
= 38,225 + 1,96(2,956848)
= 44,02042

TSP

Percentil 5 = X – 1,645δ
= 56,575– 1,645 (2,907881)
= 51,79154

Percentil 50 = 56,575

Percentil 95 = X – 1,96δ
= 56,575+ 1,96(2,907881)
= 62,27445

of all percentile calculations can be seen in table 3

Table 3. Percentile Calculation

Chair Dimensions	Anthropometric data used	Percentil (cm)			Percentil data used
		P5	P50	P95	
Chair Height	Tpo	30,62	37,18	45	P95 = 45
Chair base weight	LP	33,36	38,23	44,02	P95 = 44,02
BacK Heigt	TSP	51,79	56,58	62,27	P50 = 56,58
Chair Base Length	Ppo	38,60	44,93	52,46	P50 = 44,93
Armest Height	TSD	19,53	23,95	29,22	P50 =23,95

3.3. Chair Design

At this stage the researcher will redesign an ergonomic work chair to replace the old work chair based on body anthropometric calculations which are then used as a measure for the proposed work chair design in Figure 2. the data used are as follows

3.3.1. The seat height is 45 cm. This size is in accordance with anthropometric calculations so that it can be used for people who have long or short legs.

3.3.2. The width of the chair base is 44.5 cm. This size is in accordance with anthropometric calculations so that the seat size of the chair is wide.

3.3.3. The height of the backrest is 56.59. This size is in accordance with anthropometric calculations so that people with tall and short body sizes can use this chair.

3.3.4. The length of the base is 44.93 cm. This size is in accordance with anthropometric calculations so that people with large and small popliteal bottoms can use this chair.

3.3.5. The height of the armrests is 23.95 cm. This measure is in accordance with the anthropometric calculation of the average arm size of a person.

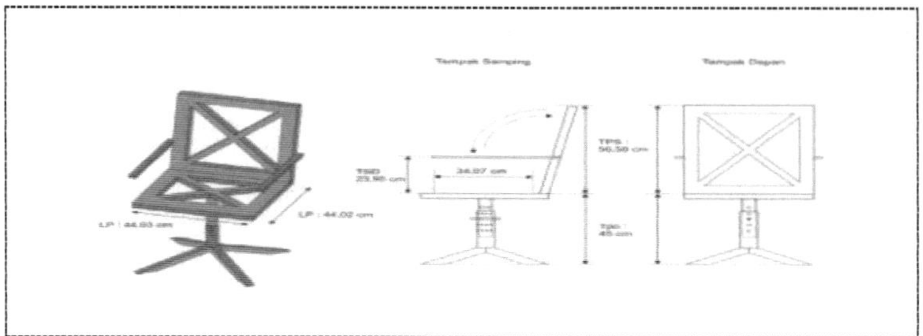

Figure 2. The Porposed framework design of chair

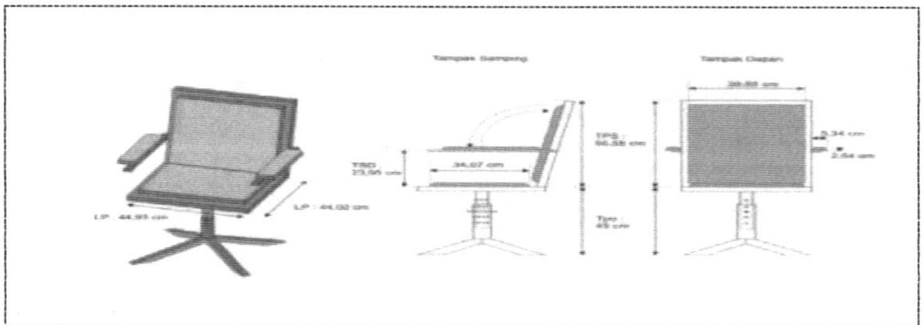

Figure 3. Porposed work chair design

References

[1] I. Dianat, J. Molenbroek, and H. I. Castellucci, "A review of the methodology and applications of anthropometry in ergonomics and product design," Ergonomics, vol. 61, no. 12, pp. 1696–1720, Dec. 2018, doi: 10.1080/00140139.2018.1502817.

[2] N. M. Namate, M. Mutingi, L. Aku, and C. Mbohwa, "Anthropometric Analysis for Ergonomic Design: Empirical Studies in Namibia," p. 8, 2017.

[3] S. O. Ismaila, A. I. Musa, S. B. Adejuyigbe, and O. D. Akinyemi, "ANTHROPOMETRIC DESIGN OF FURNITURE FOR USE IN TERTIARY INSTITUTIONS IN ABEOKUTA, SOUTH-WESTERN, NIGERIA," vol. 33, no. 3, p. 14.

[4] I. W. Taifa and D. A. Desai, "Anthropometric measurements for ergonomic design of students' furniture in India," Eng. Sci. Technol. Int. J., vol. 20, no. 1, pp. 232–239, Feb. 2017, doi: 10.1016/j.jestch.2016.08.004.

[5] A. Hedge, "Anthropometry and Workspace Design," p. 8, 2013.

[6] P. Thaneswer, J. Sanjog, A. Chowdhury, and S. Karmakar, "Applications of DHM in Agricultural Engineering: A Review," Adv. Eng. Forum, vol. 10, pp. 16–21, Dec. 2013, doi: 10.4028/www.scientific.net/AEF.10.16.

[7] E. Diniardi and A. I. Ramadhan, "Chair Design Analysis Of Work To Reduce Musculoskeletal Part 1: Anthropometry Method," vol. 5, no. 11, p. 4, 2016.

[8] J. O. Igbokwe, G. O. Osueke, U. V. Opara, M. O. Ileagu, and K. U. Ezeakaibeya, "Considerations Of Anthropometrics In The Design Of Lecture Hall Furniture," Int. J. Res. - Granthaalayah, vol. 7, no. 8, pp. 374–386, Aug. 2019, doi: 10.29121/granthaalayah.v7.i8.2019.686.

[9] S. Ansari, A. Nikpay, and S. Varmazyar, "Design and Development of an Ergonomic Chair for Students in Educational Settings," Health Scope, vol. In Press, no. In Press, Oct. 2018, doi: 10.5812/jhealthscope.60531.

[10] S. O. Ismaila, A. I. Musa, S. B. Adejuyigbe, and O. D. Akinyemi, "Anthropometric

Design Of Furniture For Use In Tertiary Institutions In Abeokuta, South- Western Nigeria," vol. 33, no. 3, p. 15.

[11] S. J., S. Karmakar, T. Patel, and A. Chowdhury, "DHM an Aid for Virtual Ergonomics of Manufacturing Shop Floor: A Review with Reference to Industrially Developing Countries," Int. J. Comput. Appl., vol. 54, no. 14, pp. 18–23, Sep. 2012, doi: 10.5120/8634-2541.

[12] A. F. Wiraputra, "Redesigning Ergonomic Vertical Folding Desk in an Apartment Room using Anthropometric Approach," p. 7, 2020.

[13]"Stephen_Pheasant_Bodyspace_Anthropometry,_Ergonomics_and_the_Design_of_the_Work,_Second_Edition__1996.pdf." .

[14] T. K. Kapuria, "Ergonomic Design of Table and Chair based on QFD and Anthropometric Measurement and improved Facility Layout," Ergon. Int. J., vol. 2, no. 3, 2018, doi: 10.23880/EOIJ-16000146.

[15] T. Patel et al., "Virtual Ergonomics Evaluation of a Design Concept of Manual Powered Portable Paddy Thresher Suitable for Hilly Region Agriculture," in Research into Design for Communities, Volume 1, vol. 65, A. Chakrabarti and D. Chakrabarti, Eds. Singapore: Springer Singapore, 2017, pp. 503–512.

A Fuzzy BWM Method for Evaluating Supplier Selection Factors in a SME Paper Manufacturer

V R B Kurniawan[1*] and F H Puspitasari[2]
[1]Department of Industrial Engineering, Universitas Sarjanawiyata Tamansiswa, Yogyakarta, Indonesia
[2]Department of Industrial Engineering, Universitas Atma Jaya Yogyakarta, Yogyakarta, Indonesia

E-mail: reza.kurniawan@ustjogja.ac.id

Abstract. In today's competitive world, companies are required to increase productivity and produce high-quality products while managing their operational efficiency. This leads a firm to involve its performing suppliers by taking into account multiple factors. For small and medium-sized enterprises (SMEs), despite its importance, supplier selection is less considered. In this study, a case study for evaluating supplier selection factors in a small-medium scale paper manufacturer in Indonesia is demonstrated using the fuzzy best-worst method (fuzzy BWM). The BWM method that provides a structurally consistent factors comparison through the best and the worst criteria is extended using a fuzzy set to address the ambiguity from decision-makers' judgement. The evaluation includes five supplier criteria: service, flexibility and delivery, reputation, quality, and purchasing cost. To determine the fuzzy criteria' weights, the nonlinearly constrained optimization problem was introduced. The result indicated that the best and the worst criteria were respectively service and purchasing cost, while the service criterion was weakly crucial than quality. Besides, the Fuzzy BWM has also provided high consistency in the comparison. The optimal weights determined in this study can be applied for ranking suppliers for further attempts.

1. Introduction

Today, product quality is significantly determined by key suppliers. Consequently, firms should consider a critical decision-making process, namely supplier selection. The right decision on selecting performing suppliers will provide many benefits for the success of a firm. The supplier selection decision is a strategic decision-making process as leading many improvement for company both reducing costs and increasing profits [1], [2], [3]. Although selecting reliable supplier is often a complex decision-making problem, the evaluation is critical to the success of the firm's overall production performance [4]. Furthermore, supplier selection affects company reputation and competitiveness in today's rapidly global changes [5].

Pertinent studies on supplier selection have majorly been solved using the multi-criteria decision-making (MCDM) approaches. The MCDM methods to solve supplier selection cases varied in types of industry both in functional organizations and in project-based organizations as conducted by Shojaei and bolvardizadeh [6]. Besides, several researchers [7], [8], [9] proposed an integrated analytical hierarchy process (AHP) and technique for order preference by similarity to ideal solution (TOPSIS) as well as extending these two integration to fuzzy set. These two methods have commonly been executed to determine criteria weights using AHP and ranking alternatives using TOPSIS, however, Junior et al. [7] compared these two methods for supplier selection demonstrated in automotive industry. In addition, Mohammed et al. [9] developed a hybrid MCDM-optimization approach to solve the sustainable supplier selection and order allocation (SSS/OA) problem in metal industry. In the study, MCDM were represented by fuzzy AHP and fuzzy TOPSIS, while the fuzzy multi-objective optimization was used to obtain the optimal order allocation.

The supplier selection problem is considered as a complex decision-making problem since involving multi-criteria and multi-alternatives. As seen in Table 1, all researchers have included the different number of criteria depending on the scope of evaluation. In the hierarchical structure of decision making, the researchers also classified the criteria either into categories or dimensions, and even deployed the criteria into sub-criteria. Several criteria were relevantly taken in many studies such as quality and cost in which these two criteria were identified, see [5], [7], [9], [10], [11], [12]. Apart from these two criteria, other criteria such as technology, agility, reputation, delivery, and relationship were also identified in several studies. Furthermore, some researchers [6], [8], [9] have included green as well as sustainability aspect into their studies.

Table 1. The different number of criteria on supplier selection evaluation identified by authors

Authors	The Criteria and The Number of Criteria
Wan et al. [10]	5 criteria and 12 sub-criteria. The criteria are quality, cost, technology, agility, and general management capability
Junior et al. [7]	6 criteria. The criteria are quality, price, delivery, supplier profile, and supplier relationship
Jain et al. [8]	3 clusters and 12 criteria. The criteria are warranties and claim policies, amount of past business, long term relationship; product development, JIT, reciprocal arrangement, packaging ability, green warehousing, carbon footprint tax, green transportation, wages, and human resource capability
Petrović et al. [5]	5 criteria. The criteria are product price, transportation costs, delivery time, company rating, and established cooperation
Shojaei and bolvardizadeh [6]	5 dimensions and 15 criteria. The criteria are similar project experience, financial/capital status, green material design, decrease of consumption for toxic materials in construction process, equipment flexibility, waste management in green construction process, dealing with plants, pollution control plans, green technologies, waste management in green construction process, dealing with plants, pollution control plans, green technologies, environmental awareness, green supports by top managers of construction companies, and green social responsibility
Fu [11]	5 criteria. The criteria are meal quality, service quality, delivery time, company image, and food safety
Mohammed et al. [9]	3 criteria and 10 sub-criteria. The criteria are conventional, green, and social.
Wang et al. [12]	5 main criteria and 15 sub-criteria. The criteria are reliability, ability, agile, effective asset management, and costs.

In this study, a case study to evaluate the supplier selection' factors in an Indonesian paper manufacturer is solved using fuzzy BWM. The extended fuzzy BWM is proposed as providing a structurally consistent factor comparison, while the fuzzy environment models a more real MCDM case. Previously, supplier selection studies using BWM has been carried out by Gupta and Barua [13],

Badi and Ballem [14], and Cheraghalipour et al. [15], while Wu et al. [16] and Amiri et al. [17] proposed the fuzzy BWM. Although BWM was integrated with other MCDM methods such as fuzzy TOPSIS, MAIRCA, and VIKOR [13], [14], [15], the method was deliberately used to obtain criteria weights, while other methods were chosen to rank the best supplier. Meanwhile, Wu et al. [16] and Amiri et al. [17] revealed that fuzzy-based MCDM had several advantages in complex modelling and precise data collecting, moreover, Wu et al. [16] extended to the interval type-2 fuzzy sets. Subsequently, it is important to evaluate suppliers in small and medium-sized enterprises (SMEs) considering that SMEs pay close attention to the quality of their products in order to compete with large enterprises and this refers to the important supplier's role. As stated by Rezaei et al. [18] that An SME is suggested to build strategic relationships with its suppliers even in particular functional areas. On the other hand, due to lack of awareness and competencies in decision-making practices, the supplier evalution in SME is often less considered. Therefore, the decision factors' rank on supplier selection determined in this study can be used as a guide for decision-makers (DMs), especially for SMEs located in developing countries, to choose suitable suppliers taking into account the priority factors.

2. Method

BWM was firstly introduced by Rezaei [19] and fuzzy BWM was developed by Guo and Zhao [20]. For the initial step, the DMs determine the decision goal and identify the criteria to evaluate a decision-making problem. This step will generate n criteria (C_1, C_2, ..., C_n) for supplier selection decision-making process. Secondly, the DMs are asked to determine the best and the worst criteria among the n decision criteria. The best and the worst criteria are represented with C_B and C_W, respectively. In the first two steps, the fuzzy BWM is executed in the same way as the BWM.

Then, fuzzy reference comparisons are executed using triangular fuzzy number. The triangular fuzzy numbers are obtained based on the transformation rules of linguistic variables, which are: equally important (EI), weakly important (WI), fairly important (FI), very important (CI), and absolutely important (AI) in which the membership functions are respectively (1,1,1), (2/3, 1, 3/2), (3/2, 2, 5/2), (5/2, 3, 7/2), and (7/2, 4, 9/2) [20]. The fuzzy reference comparison for the best criterion produces the fuzzy Best-to-Others (BO) vector $\tilde{A}_B = (\tilde{a}_{B1}, \tilde{a}_{B2}, \ldots, \tilde{a}_{Bn})$, while fuzzy Others-to-Worst (OW) vector $\tilde{A}_W = (\tilde{a}_{1W}, \tilde{a}_{2W}, \ldots, \tilde{a}_{nW})$ is indicated by the fuzzy reference comparison for the worst criterion. To obtain optimal criteria weights, the nonlinearly constrained optimization problem is build as indicated in equation (1) [20].

- From the **File** menu select the **Page Setup** … option to display the *Page Setup* dialog box.
- Select the paper size you want to use (US Letter or A4).
- Depending on the paper size you use, adjust the margin settings to those shown in Table 1.

$$\min \xi$$
$$s.t.\begin{cases} \left|\dfrac{\tilde{w}_B}{\tilde{w}_j} - \tilde{a}_{Bj}\right| \leq \xi \\ \left|\dfrac{\tilde{w}_j}{\tilde{w}_W} - \tilde{a}_{jW}\right| \leq \xi \\ \sum_{j=1}^{n} R(\tilde{w}_j) = 1 \\ l_j^w \leq m_j^w \leq u_j^w \\ l_j^w \geq 0 \\ j = 1,2,\ldots,n \end{cases} \tag{1}$$

Suppose that $\xi^* = (k^*, k^*, k^*)$, then the equation (1) can be transferred as

$$\min \xi$$
$$s.t. \begin{cases} \left| \frac{(l_B^w, m_B^w, u_B^w)}{(l_j^w, m_j^w, u_j^w)} - (l_{Bj}, m_{Bj}, u_{Bj}) \right| \leq (k^*, k^*, k^*) \\ \left| \frac{(l_j^w, m_j^w, u_j^w)}{(l_W^w, m_W^w, u_W^w)} - (l_{jW}, m_{jW}, u_{jW}) \right| \leq (k^*, k^*, k^*) \\ \sum_{j=1}^{n} R(\widetilde{w}_j) = 1 \\ l_j^w \leq m_j^w \leq u_j^w \\ l_j^w \geq 0 \\ j = 1,2, \ldots, n \end{cases} \quad (2)$$

$$s.t. \begin{cases} \left| \frac{(l_B^w, m_B^w, u_B^w)}{(l_j^w, m_j^w, u_j^w)} - (l_{Bj}, m_{Bj}, u_{Bj}) \right| \leq (k^*, k^*, k^*) \\ \left| \frac{(l_j^w, m_j^w, u_j^w)}{(l_W^w, m_W^w, u_W^w)} - (l_{jW}, m_{jW}, u_{jW}) \right| \leq (k^*, k^*, k^*) \\ \sum_{j=1}^{n} R(\widetilde{w}_j) = 1 \\ l_j^w \leq m_j^w \leq u_j^w \\ l_j^w \geq 0 \\ j = 1,2, \ldots, n \end{cases}$$

Finally, consistency checking is executed to ensure the consistency level of pairwise comparison. The consistency value is indicated by the value of consistency ratio (CR) which is generated from the value of ξ^* is divided by the value of consistency index (CI). The CI value refers to $\tilde{a}_{BW}$ for EI (3.00), WI (3.80), FI (5.29), VI (6.69), and AI (8.04) [20].

3. Result and Discussion

3.1. Building the decision criteria system

Initially, a set of decision criteria is identified by inviting three DMs in the enterprise who are procurement admin, process manager, and product specialist. The three DMs agreed to consider five relevant criteria used in supplier selection decision, which are: service (C_1), flexibility and delivery (C_2), reputation (C_3), quality (C_4), and purchasing cost (C_5). These five criteria are factors that are often considered by the management during its operation.

3.2. Calculating fuzzy comparison based on the best and the worst criteria

The best and the worst criteria are subsequently determined by the DMs that in this case, the best criterion is service (C_1), while purchasing cost (C_5) becomes the worst criterion. Then, the fuzzy pairwise comparison is generated taking into account the DMs' opinion when comparing between two criteria. The fuzzy preferences between the best criterion, C_1, and the worst criterion, C_2, is absolutely important, while the other best and worst preferences as shown in Table 2 and Table 3.

Table 2. The linguistic terms for fuzzy preference of the best criterion

Criteria	C_1	C_2	C_3	C_4	C_5
Best criterion C_1	EI	FI	VI	WI	AI

Table 3. The linguistic terms for fuzzy preference of the worst criterion

Criteria	Worst criteria C_5
C_1	AI
C_2	FI
C_3	WI
C_4	VI
C_5	EI

The fuzzy reference comparisons for the best and the worst criteria are obtained based on the fuzzy preference as described in Table 2 and Table 3 and represented by the fuzzy BO vector, $\tilde{A}_B$, and the fuzzy OW vector, $\tilde{A}_W$, as follows

$$\tilde{A}_B = [(1,1,1), \left(\frac{3}{2}, 2, \frac{5}{2}\right), \left(\frac{5}{2}, 3, \frac{7}{2}\right), \left(\frac{2}{3}, 1, \frac{3}{2}\right), \left(\frac{7}{2}, 4, \frac{9}{2}\right)$$

$$\tilde{A}_W = [\left(\frac{7}{2}, 4, \frac{9}{2}\right), \left(\frac{3}{2}, 2, \frac{5}{2}\right), \left(\frac{2}{3}, 1, \frac{3}{2}\right), \left(\frac{5}{2}, 3, \frac{7}{2}\right), (1,1,1)]$$

3.3. Finding optimal weights through the optimization problem

To find the optimal weights for the five criteria, the constrained optimization problem is introduced as follows

$$\min \tilde{\xi}^*$$

$$s.t. \begin{cases} \left|\frac{(l_1^W, m_1^W, u_1^W)}{(l_2^W, m_2^W, u_2^W)} - (l_{12}, m_{12}, u_{12})\right| \le (k^*, k^*, k^*) \\ \left|\frac{(l_1^W, m_1^W, u_1^W)}{(l_3^W, m_3^W, u_3^W)} - (l_{13}, m_{13}, u_{13})\right| \le (k^*, k^*, k^*) \\ \left|\frac{(l_1^W, m_1^W, u_1^W)}{(l_4^W, m_4^W, u_4^W)} - (l_{14}, m_{14}, u_{14})\right| \le (k^*, k^*, k^*) \\ \left|\frac{(l_1^W, m_1^W, u_1^W)}{(l_5^W, m_5^W, u_5^W)} - (l_{15}, m_{15}, u_{15})\right| \le (k^*, k^*, k^*) \\ \left|\frac{(l_2^W, m_2^W, u_2^W)}{(l_5^W, m_5^W, u_5^W)} - (l_{25}, m_{25}, u_{25})\right| \le (k^*, k^*, k^*) \\ \left|\frac{(l_3^W, m_3^W, u_3^W)}{(l_5^W, m_5^W, u_5^W)} - (l_{35}, m_{35}, u_{35})\right| \le (k^*, k^*, k^*) \\ \left|\frac{(l_4^W, m_4^W, u_4^W)}{(l_5^W, m_5^W, u_5^W)} - (l_{45}, m_{45}, u_{45})\right| \le (k^*, k^*, k^*) \\ \sum_{j=1}^{5} R(\tilde{w}_j) = 1 \\ l_j^W \le m_j^W \le u_j^W;\ l_j^W \ge 0 \\ j = 1, \dots, 5 \end{cases} \quad (3)$$

The nonlinearly constrained optimization problem can be built as follows

$$\min k^* \quad (4)$$

IOP Conf. Series: Materials Science and Engineering 1071 (2021) 012004 doi:10.1088/1757-899X/1071/1/012004

$$
s.t.\begin{cases}
l_1-\frac{3}{2}*u_2\le k*u_2;l_1-\frac{3}{2}*u_2\ge -k*u_2;\\
u_1-\frac{5}{2}*l_2\le k*l_2;u_1-\frac{5}{2}*l_2\ge -k*l_2;\\
l_1-\frac{5}{2}*u_3\le k*u_3;l_1-\frac{5}{2}*u_3\ge -k*u_3;\\
m_1-3*m_3\le k*m_3;m_1-3*m_3\ge -k*m_3;\\
u_1-\frac{7}{2}*l_3\le k*l_3;u_1-\frac{7}{2}*l_3\ge -k*l_3;\\
l_1-\frac{2}{3}*u_4\le k*u_4;l_1-\frac{2}{3}*u_4\ge -k*u_4;\\
m_1-1*m_4\le k*m_4;m_1-1*m_4\ge -k*m_4;\\
u_1-\frac{3}{2}*l_4\le k*l_4;u_1-\frac{3}{2}*l_4\ge -k*l_4;\\
l_1-\frac{7}{2}*u_5\le k*u_5;l_1-\frac{7}{2}*u_5\ge -k*u_5;\\
m_1-4*m_5\le k*m_5;m_1-4*m_5\ge -k*m_5;\\
u_1-\frac{9}{2}*l_5\le k*l_5;u_1-\frac{9}{2}*l_5\ge -k*l_5;\\
l_2-\frac{3}{2}*u_5\le k*u_5;l_2-\frac{3}{2}*u_5\ge -k*u_5;\\
m_2-2*m_5\le k*m_5;m_2-2*m_5\ge -k*m_5;\\
u_2-\frac{5}{2}*l_5\le k*l_5;u_2-\frac{5}{2}*l_5\ge -k*l_5;\\
l_3-\frac{2}{3}*u_5\le k*u_5;l_3-\frac{2}{3}*u_5\ge -k*u_5;\\
m_3-1*m_5\le k*m_5;m_3-1*m_5\ge -k*m_5;\\
u_3-\frac{3}{2}*l_5\le k*l_5;u_3-\frac{3}{2}*l_5\ge -k*l_5;\\
l_4-\frac{5}{2}*u_5\le k*u_5;l_4-\frac{5}{2}*u_5\ge -k*u_5;\\
m_4-3*m_5\le k*m_5;m_4-3*m_5\ge -k*m_5;\\
u_4-\frac{7}{2}*l_5\le k*l_5;u_4-\frac{7}{2}*l_5\ge -k*l_5;\\
\frac{1}{6}*l_1+\frac{1}{6}*4*m_1+\frac{1}{6}*u_1+\frac{1}{6}*l_2+\frac{1}{6}*4*m_2+\frac{1}{6}*u_2+\\
\frac{1}{6}*l_3+\frac{1}{6}*4*m_3+\frac{1}{6}*u_3+\frac{1}{6}*l_4+\frac{1}{6}*4*m_4+\frac{1}{6}*u_4+\\
\frac{1}{6}*l_5+\frac{1}{6}*4*m_5+\frac{1}{6}*u_5=1\\
l_1\le m_1\le u_1;\ l_2\le m_2\le u_2;\ l_3\le m_3\le u_3;\\
l_4\le m_4\le u_4;\ l_5\le m_5\le u_5;\\
l_1\ge 0;l_2\ge 0;l_3\ge 0;l_4\ge 0;l_5\ge 0
\end{cases}
$$

By solving equation (4), the optimal fuzzy weights is represented in Figure 1. The obtained optimal fuzzy weights for the five criteria namely service ($\widetilde{w}_1^*$), flexibility and delivery ($\widetilde{w}_2^*$), reputation ($\widetilde{w}_3^*$), quality ($\widetilde{w}_4^*$), and purchasing cost ($\widetilde{w}_5^*$), which are:

$\widetilde{w}_1^*=(0.261,0.330,0.366)$; $\widetilde{w}_2^*=(0.161,0.184,0.199)$; $\widetilde{w}_3^*=(0.094,0.119,0.119)$
$\widetilde{w}_4^*-(0.250,0.293,0.335)$; $\widetilde{w}_5^*-(0.085,0.085,0.085)$
$\tilde{\xi}^*=(0.037,0.037,0.037)$

Then, the crisp weights or so-called as graded mean integration representation (GMIR) can be calculated by $(l_j^w+4m_j^w+u_j^w)/6$. As a result, the weights for service (w_1^*), flexibility and delivery

(w_2^*), reputation (w_3^*), quality (w_4^*), and purchasing cost (w_5^*) are 0.325, 0.182, 0.115, 0.293, and 0.085, respectively.

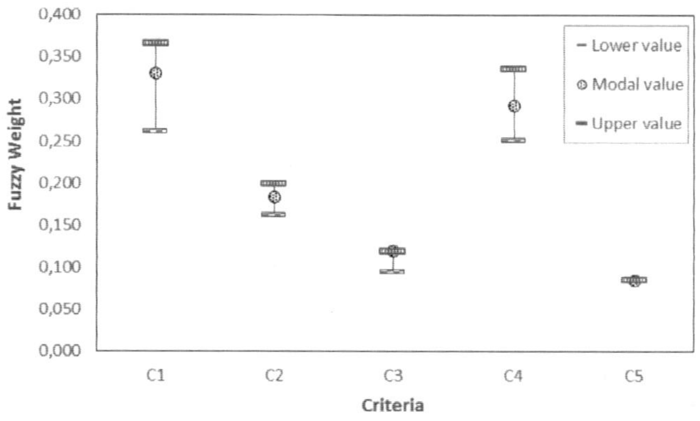

Figure 1. The optimal fuzzy weights for the five criteria

The priority order of supplier selection' criteria is C_1>C_4>C_2>C_3>C_5. The optimal weights for the five criteria is obtained with the consistency index (CI) of 8.04 ($a_{BW} = a_{15}$ = (7/2, 4, 9/2)) and the consistency ratio (CR) of 0.037/8.04 = 0.0046. A very high consistency is indicated since the CR value is almost zero. The result also indicates that the service criterion (C_1) is the most important factor as representing the main supplier's indicators expected by the company such as responsiveness to solve issues, providing easy access for orders, and the ability to provide information required. Besides, in this case, quality criterion (C_4) is also considered as top priority factor after service criteria. The C_4 criterion includes the consistency of supplier in providing quality materials which guarantee the standardized finished goods. By contrast, purchasing cost (C_5) is the least important factors since the cost may not affect the overall supplier performances, further, the issues related to cost such as payment methods, price fluctuations and other supply chain costs are negotiable solved between two parties.

The criteria order defined is relevant to the company's current situation. Service performance and quality are the key factors in building long-term relationship with suppliers considering that SMEs are vulnerable for changes due to insufficient practices. It is also supported by Wan et al. [10] which included response to change and ease of communication in the service performance category. In this paper manufacturer, the suppliers are expected to be responsive and accessible. Then, the second factor is quality which comprises of rejection rate of product, quality management, and after sales service quality [7], [10]. In this case, the quality of supplier will determine the quality of material produced and the after sales service is closely related to the service criterion. The other two criteria, C_2 and C_3, are grouped as the secondary priority since these two criteria appear as a consequence of the first two criteria. For example, the reputation criterion appears as an effect that the supplier performs good long-term relationship with the company due to its reliable service and quality. Therefore, by considering this priority factors, the DMs in the company are able to consider the driving factors particularly in selection suppliers based on their specifications.

4. Conclusion

In this SME paper manufacturer, service (C_1) and purchasing cost (C_5) are the best criterion and the worst criterion, respectively. Besides, the optimal weights for five criteria are obtained with a very high consistent result indicated by CR value. The fuzzy BWM proposed in this study solves the DMs' judgement ambiguity in a structured comparison. To achieve deep evaluation on supplier selection, the further decision-making process with alternatives can be conducted using other MCDM techniques as well as investigating the causal relations among criteria.

References

[1] E. A. Frej, L. R. P. Roselli, J. Araújo De Almeida, and A. T. De Almeida, "A Multicriteria Decision Model for Supplier Selection in a Food Industry Based on FITradeoff Method," *Math. Probl. Eng.*, vol. 2017, 2017.

[2] K. Nourianfar and G. A. Montazer, "A fuzzy MCDM approach based on COPRAS method to solve supplier selection problems," *IKT 2013 - 2013 5th Conf. Inf. Knowl. Technol.*, pp. 231–235, 2013.

[3] V. Agrawal, A. M. Agrawal, and R. P. Mohanty, "Application of fuzzy MCDM in supplier selection of fertiliser manufacturing industry," *Int. J. Bus. Perform. Supply Chain Model.*, vol. 9, no. 2, pp. 133–159, 2017.

[4] A. Yıldız and A. Y. Yayla, "MULTI-CRITERIA DECISION-MAKING METHODS FOR SUPPLIER SELECTION: A LITERATURE REVIEW," *South African J. Ind. Eng. August*, vol. 26, no. 2, pp. 319–323, 2015.

[5] G. Petrović, J. Mihajlović, Ž. Ćojbašić, M. Madić, and D. Marinković, "Comparison of three fuzzy MCDM methods for solving the supplier selection problem," *Facta Univ. Ser. Mech. Eng.*, vol. 17, no. 3, pp. 455–469, 2019.

[6] P. Shojaei and A. bolvardizadeh, "Rough MCDM model for green supplier selection in Iran: a case of university construction project," *Built Environ. Proj. Asset Manag.*, vol. 10, no. 3, pp. 437–452, 2020.

[7] F. R. Lima Junior, L. Osiro, and L. C. R. Carpinetti, "A comparison between Fuzzy AHP and Fuzzy TOPSIS methods to supplier selection," *Appl. Soft Comput. J.*, vol. 21, pp. 194–209, 2014.

[8] N. Jain, A. R. Singh, and R. K. Upadhyay, "Sustainable supplier selection under attractive criteria through FIS and integrated fuzzy MCDM techniques," *Int. J. Sustain. Eng.*, vol. 00, no. 00, pp. 1–22, 2020.

[9] A. Mohammed, I. Harris, and K. Govindan, "A hybrid MCDM-FMOO approach for sustainable supplier selection and order allocation," *Int. J. Prod. Econ.*, vol. 217, no. February, pp. 171–184, 2019.

[10] S. ping Wan, G. li Xu, and J. ying Dong, "Supplier selection using ANP and ELECTRE II in interval 2-tuple linguistic environment," *Inf. Sci. (Ny).*, vol. 385–386, pp. 19–38, 2017.

[11] Y. K. Fu, "An integrated approach to catering supplier selection using AHP-ARAS-MCGP methodology," *J. Air Transp. Manag.*, vol. 75, no. July 2018, pp. 164–169, 2019.

[12] C. N. Wang, H. T. Tsai, T. P. Ho, V. T. Nguyen, and Y. F. Huang, "Multi-criteria decision making (MCDM) model for supplier evaluation and selection for oil production projects in vietnam," *Processes*, vol. 8, no. 2, 2020.

[13] H. Gupta and M. K. Barua, "Supplier selection among SMEs on the basis of their green innovation ability using BWM and fuzzy TOPSIS," *J. Clean. Prod.*, vol. 152, pp. 242–258, 2017.

[14] I. Badi and M. Ballem, "Supplier Selection using Rough BWM-MAIRCA model: A case study in Pharmaceutical Supplying in Libya," *Decis. Mak. Appl. Manag. Eng.*, vol. 1, no. 2, pp. 16–33, 2018.

[15] A. Cheraghalipour, M. M. Paydar, and M. Hajiaghaei-Keshteli, "Applying a hybrid BWM-VIKOR approach to supplier selection: A case study in the Iranian agricultural implements industry," *Int. J. Appl. Decis. Sci.*, vol. 11, no. 3, pp. 274–301, 2018.

[16] Q. Wu, L. Zhou, Y. Chen, and H. Chen, "An integrated approach to green supplier selection based on the interval type-2 fuzzy best-worst and extended VIKOR methods," *Inf. Sci. (Ny).*, vol. 502, pp. 394–417, 2019.

[17] M. Amiri, M. Hashemi-Tabatabaei, M. Ghahremanloo, M. Keshavarz-Ghorabaee, E. K. Zavadskas, and A. Banaitis, "A new fuzzy BWM approach for evaluating and selecting a sustainable supplier in supply chain management," *Int. J. Sustain. Dev. World Ecol.*, vol. 00, no. 00, pp. 1–18, 2020.

[18] J. Rezaei, R. Ortt, and P. Trott, "How SMEs can benefit from supply chain partnerships," *Int. J. Prod. Res.*, vol. 53, no. 5, pp. 1527–1543, 2015.
[19] J. Rezaei, "Best-worst multi-criteria decision-making method," *Omega (United Kingdom)*, vol. 53, pp. 49–57, 2015.
[20] S. Guo and H. Zhao, "Fuzzy best-worst multi-criteria decision-making method and its applications," *Knowledge-Based Syst.*, vol. 121, pp. 23–31, 2017.

IOP Conf. Series: Materials Science and Engineering 1071 (2021) 012005 doi:10.1088/1757-899X/1071/1/012005

Designing Optimal Distribution Routes using a Vehicle Routing Problem (VRP) Model in a Logistics Service Provider

F H Puspitasari[1*] and V R B Kurniawan[2]
[1]Department of Industrial Engineering, Universitas Atma Jaya Yogyakarta, Yogyakarta, Indonesia
[2]Department of Industrial Engineering, Universitas Sarjanawiyata Tamansiswa, Yogyakarta, Indonesia

E-mail: fransiska.hernina@uajy.ac.id

Abstract. In a running business, it is common for companies to face some challenges. One of them is related to supply chain performances, especially product distribution. Companies often have a problem determining the distribution route from the origin (warehouse) to the destination (consumer). This research will discuss the distribution route in a case study at a logistics service provider in Yogyakarta. The problem is that each courier does not have a fixed distribution route, causing couriers to overload and less effective when distributing goods to consumers. To solve this problem, we created a delivery zone for each courier done in previous studies using courier assignment optimization to minimize mileage. Furthermore, based on the zones that have been created, we determined the route using the classical vehicle routing problem (VRP) mathematical model to minimize travel time in this study. Based on the results obtained, couriers not only reduce mileage but also travel time, which makes this distribution route more optimal. Later, a vehicle routing problem with a time window can be carried out and compared with the current VRP for further research.

1. Introduction

Supply Chain Management (SCM) is a crucial element when a company runs its business system. Why? When the company has managed its supply chain well, consumer demand can be ensured to be met, or the service level at the company is close to 100%. However, indeed, this is very problematic. Many things can be a challenge to supply chain performance.

According to Li [1], one of the supply chain challenges is the configuration process in the supply chain network and the arrangement of the goods transportation process to minimize production, inventory, and transportation costs. Even a process of distributing products from place of origin to a destination can be a significant problem affecting supply chain performance [2]. Therefore, it is a manager's job to design a distribution route [3].

Much of the current literature on some journals pay particular attention to a distribution problem, especially determining a case study distribution route. There are some cases in waste recycling industries about waste management. Here, vehicles will collect and/or drop-off the waste at specific points, taking into account vehicles' capacity and minimization of total travel distances [4]–[6]. Nowakowski [4] took a case study in Poland to collect WEEE (waste electrical and electronic equipment) and put them into a recycling company. The route is gained using a combination of CVRP (capacitated vehicle routing problem) and VRLP (vehicle routing problem). Kuo [5] solved a problem in Indonesia to pick-up garbage from temporary dumps and drop-off them to the main dump applying a combination of CVRP, Genetic Algorithm, Ant Colony Optimization, and two-local search algorithm. Puspita [6] has a similar problem as Kuo [5], but the study case took place in Palembang, Indonesia, using a mathematical model of RC-OCVRP (robust counterpart open capacitated vehicle routing).

Other case studies have also examined the route distribution problem to minimize travel distances in some other industries. It is because distribution is the main problem that can affect business performance. Mingprasert [7] found a distribution problem in the plastic packaging industry. The industry had received many complaints from its customers due to delivery delays. Thus, it urged the industry to solve this problem by optimizing the distribution routes using a combination of multiple capacitated vehicle routing problems and Artificial Bee Colony and the objective of minimizing total traveling distance [7]. Another case occurred in a natural gas company. It applied the CVRP and Saving Algorithm to determine the natural gas distribution regarding time windows [8]. Also, a study case in a supermarket chain in which an effective and efficient distribution route is crucial. Comert [9] tried for optimizing a route distribution using the Clustering Algorithm and Branch-and-Bound Algorithm. Here, the customers were grouped applying the Clustering Algorithm, and every group is determined the route using the Branch-and-Bound Algorithm.

Two other studies are in service industries, hospitals, and a post-office. Clarke [10] optimized some distribution routes for supplying medications to rural and remote hospitals. Here, it worked with the concept of pick-ups and drop-offs. The vehicles collected the medications from some pick-up areas and would drop-off them in the hospitals [10]. Clarke [10] used CVRP with synchronous pickups-dropoffs to minimize the total distances and costs. Sbai [11] solved distribution routes in Tunisian Post Office. They determined the delivery routes for mail and parcel pick-up to minimize total distances using CVRP with HGA-VNS (genetic variable neighborhood search algorithm).

Overall, these case studies highlight the need to minimize total travel distances, whereas we can also minimize transportation costs or travel time. This paper will determine several distribution routes in particular regions that minimize the total travel distances and travel time in a logistic service provider in Yogyakarta. Given all that has been mentioned so far, most of them focus on minimizing total travel distances. One paper mentioned that it grouped customers first then made the distribution route. This research also re-groups the customers or regions using a modified assignment problem to minimize total travel distances, which it had been carried out [12]. This paper will then determine the optimal routes using VRP classic with Branch-and-Bound Algorithm to minimize total travel time.

2. Method

The main objective is to find an optimum route for each courier in the city of Yogyakarta. The first method is to group the regions for each courier using a modified assignment problem model [12]. This model aims to minimize the total mileage of each courier by paying attention to capacity. After grouping the regions, each courier's delivery route is determined with the aim of minimizing the total travel time in each courier region using a vehicle routing problem (VRP) model and Branch-and-Bound Algorithm. In this model, every region must be visited once and must be met with the demands. The notations and model are presented below.

Notations:

N_c = a set of regions (N_1, N_2, N_3, ...)
N = a set of the warehouse and regions (N_0, N_1, N_2, ...)
x_{ij} = 1, if there is a route from i to j; 0, otherwise
y_{ij} = number of units distributed from i to j.
c_{ij} = travel time from i to j.
D_i = a demand for region i.

$$\text{Minimize } Z = \sum_{i \in N} \sum_{j \in N} c_{ij} \times x_{ij} \quad (1)$$

s.t.

$$\sum_{i\in N} x_{ip} = \sum_{j\in N} x_{pj} \quad , \forall p \in N_c \tag{2}$$

$$\sum_{i\in N} x_{ij} = 1 \quad , \forall j \in N_c \tag{3}$$

$$\sum_{j\in N} y_{ij} - \sum_{j\in N} y_{ji} = D_i \quad , \forall j \in N_c \tag{4}$$

$$D_i \times x_{ij} \le y_{ij} \quad , \forall i, j \in N \tag{5}$$

$$x_{ij} \in \{0,1\} \tag{6}$$

To minimize the total travel time in each zone, equation (1) is calculated by multiplying travel time from the warehouse to regions with the number of quantities delivered. Equation (2) shows a vehicle visiting a region *i* must leave it immediately. Equation (3) indicates a region *i* should only be visited once. Equations (4) and (5) indicate that quantities distributed must be according to the demands. The last equation (6) points out a decision variable that must be a binary variable, which 1 shows a route from region *i* to *j*, 0 otherwise.

3. Results and Discussions

A logistics service provider serves good deliveries to areas in the city of Yogyakarta. It owns one big warehouse as an origin point storing many goods. Once packages arrive in the warehouse, it will be delivered to the ultimate customers in a region. It provides the deliveries which are not only in Daerah Istimewa Yogyakarta but also in Klaten and Magelang. However, this case study is only limited to areas in the city of Yogyakarta, having 45 regions served and divided into four zones. We focus on these regions because they are thickly crowded based on a data history of delivery volume and frequency. Also, there is an overload problem here. Existing distribution routes do not exist. In other words, a courier defines the route randomly, which may cause ineffectively and inefficient routes. At the same time, it has another problem that the courier has a limited capacity.

To solve this problem, we re-group the existing zones to minimize the total distances, taking into account each courier's limited capacity, 55 kgs, and demands that must be met. Zone 1 was overcapacity that led to Zone 2, 3, and 4 having uneven capacity distribution from the previous zones. Using the modified assignment optimization model, we can overcome the overcapacity problem, and the capacity distribution for each zone becomes more evenly [12]. Despite being still four zones, each new zone consists of different regions [12]. It is shown in Table 1.

Table 1. The new zones with the optimal condition [12]

Region	**Avg daily vol (kg)**	**Region**	**Avg daily vol (kg)**	**Region**	**Avg daily vol (kg)**	**Region**	**Avg daily vol (kg)**
Zone 1		Zone 2		Zone 3		Zone 4	
SRMJ$_{(DR)}$	3.81	PGKM$_{(GT)}$	2.26	TGPG$_{(DR)}$	1.59	BCRO$_{(GK)}$	6.19
DMGN$_{(GK)}$	7.37	SSRM$_{(GT)}$	3.56	BSRN$_{(DR)}$	4.96	MNTR$_{(MR)}$	4.93
GOWN$_{(JT)}$	11.15	KTBR$_{(GK)}$	3.93	TRBN$_{(GK)}$	7.70	KPRN$_{(MG)}$	3.33
TGRJ$_{(TR)}$	23.00	KTRN$_{(GK)}$	6.59	PRGN$_{(KG)}$	1.52	NMPN$_{(NP)}$	1.93
SRST$_{(UB)}$	8.52	PWDJ$_{(GM)}$	2.70	PRBY$_{(KG)}$	0.30	GNKT$_{(PL)}$	0.70
		NPSN$_{(GM)}$	7.11	KDPT$_{(KR)}$	1.15	KRCK$_{(TR)}$	2.44
		BMJO$_{(JT)}$	2.11	BTKM$_{(MG)}$	4.74	KRWR$_{(TR)}$	4.37
		CKRD$_{(JT)}$	4.37	NPRJ$_{(NP)}$	3.22	SMKI$_{(UB)}$	1.56
		RJWN$_{(KG)}$	4.67	PWKT$_{(PL)}$	4.30	WRBT$_{(UB)}$	6.19
		PTHN$_{(KR)}$	1.63	BNER$_{(TR)}$	15.44	PNDY$_{(UB)}$	4.96

		$PNMB_{(KR)}$	3.26	$MJMJ_{(UB)}$	9.56	$GWNG_{(UB)}$	2.67
		$SRYD_{(MR)}$	5.19			$THNN_{(UB)}$	4.85
		$GDKW_{(MR)}$	3.22			$WRBJ_{(WB)}$	2.89
		$WRGN_{(MG)}$	4.33			$PTPL_{(WB)}$	2.48
						$PKCN_{(WB)}$	4.81
$\sum$	**53.85**		**54.93**		**54.5**		**54.3**

Having defined the optimal zones, we will now move on to determine a distribution route in Zone 1, 2, 3, and 4. Firstly, time travel matrices in each zone are obtained with the origin started from the warehouse. The matrices assumed symmetric are the costs, c_{ij}, which will be minimized in the objective function. Also, we consider demand in every region that must be fulfilled. Zone 1 has five regions that must be visited with a total demand of 53.85 kgs out of 55 kgs. Zone 2 has 14 regions that must be visited with a total demand of 54.93 kgs out of 55 kgs. Zone 3 has 11 regions that must be visited with a total demand of 54.5 kgs out of 55 kgs. Zone 4 has 15 regions that must be visited with a total demand of 54.3 kgs out of 55 kgs. Each region must be traveled once, as revealed in the mathematical model earlier. Working with the data and using LINGO 18.0, we calculate and obtain four zones' distribution routes, as can be seen from Figure 1 – 4.

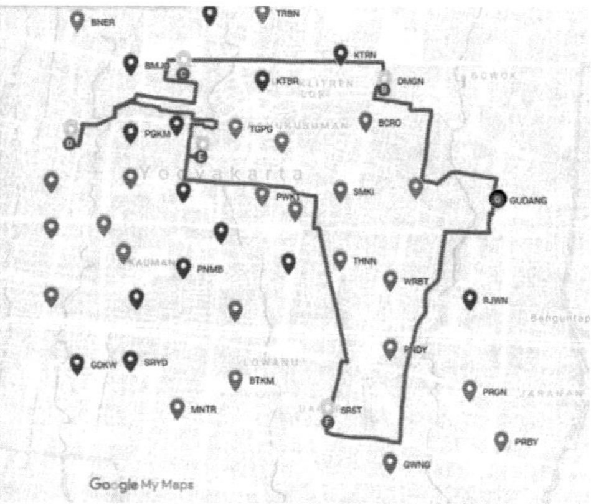

Figure 1. A distribution route of Zone 1

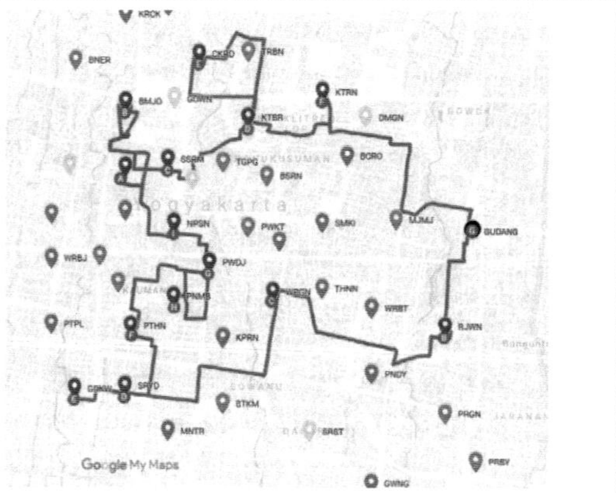

Figure 2. A distribution route of Zone 2

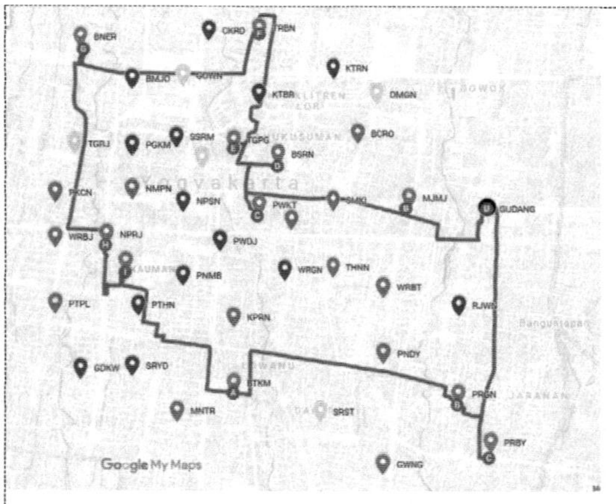

Figure 3. A distribution route of Zone 3

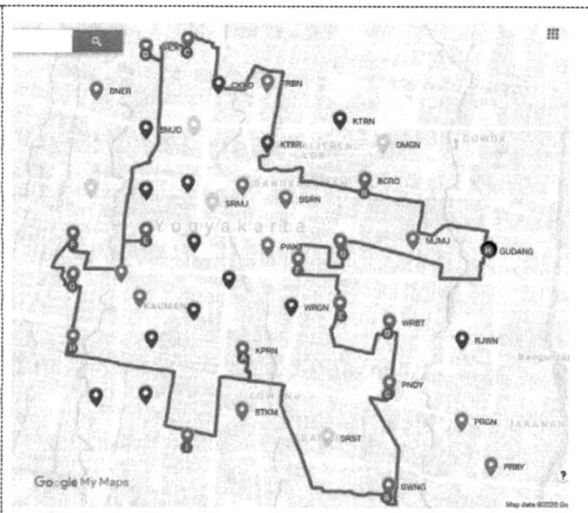

Figure 4. A distribution route of Zone 4

From the figures above, we can see that the distribution route of each zone. Zone 1 is started from the warehouse, then goes to DMG, GOWN, TGRJ, SRMJ, SRST, and comes back to the warehouse, which needs 58 minutes to travel. Zone 2 is started from the warehouse then heads to RJWN, WRGN, SRYD, GDKW, PTHN, PWDJ, PNMB, NPSN, PGKM, BMJO, SSRM, KTBR, CKRD, KTRN, and returns to the warehouse, which needs 97 minutes to travel. Zone 3 is started from the warehouse then goes to MJMJ, PWKT, BSRN, TGPG, TRBN, BNER, NPRJ, KDPT, BTKM, PRGN, PRBY, then comes back to the warehouse, which needs 86 minutes to travel. Zone 4 is begun from the warehouse then heads to BCRO, KRWR, KRCK, NMPN, PKCN, WRBJ, PTPL, MNTR, KPRN, GWNG, PNDY, WRBT, THNN, GNKT, SMKI, and heads back to the warehouse, which required 97 minutes.

4. Conclusions

In this investigation, the aim was to determine the optimal routes of four zones by minimizing total travel time. After defining the new four zones taking into account the mileage minimization in the previous research, a distribution route is made for every zone. Zone 1 consists of 5 regions and has a total travel time of 58 minutes. Zone 2 consists of 14 regions and has a total travel time of 97 minutes. Zone 3 consists of 11 regions and has a total travel time of 86 minutes. Zone 4 consists of 15 regions and has a total travel time of 97 minutes. Further research in this field would be of great help in using time windows or considering road congestion.

References

[1] L. Li, *Managing Supply Chain and Logistics: Competitive Strategy for A Sustainable Future*. Singapore: World Scientific Publishing Co. Pte. Ltd., 2014.

[2] F. R. Jacobs and R. B. Chase, *Operatons and Supply Chain Management*. UK: McGraw Hill, 2014.

[3] M. . Hugos, *Essentials of Supply Chain Management*. New Jersey: John Wiley & Sons, Inc, 2003.

[4] P. Nowakowski, “A proposal to improve e-waste collection efficiency in urban mining: Container loading and vehicle routing problems – A case study of Poland,” *Waste Manag.*, vol. 60, pp. 494–504, 2017, doi: 10.1016/j.wasman.2016.10.016.

[5] R. J. Kuo and F. E. Zulvia, “Hybrid genetic ant colony optimization algorithm for capacitated vehicle routing problem with fuzzy demand - A case study on garbage collection system,” *2017 4th Int. Conf. Ind. Eng. Appl. ICIEA 2017*, pp. 244–248, 2017, doi: 10.1109/IEA.2017.7939215.

[6] F. M. Puspita, Y. Hartono, N. Z. Syaputri, E. Yuliza, and W. D. Pratiwi, “Robust Counterpart Open Capacitated Vehicle Routing (RC-OCVRP) Model in Optimization of Garbage Transportation in District Sako and Sukarami, Palembang City,” *Int. J. Electr. Comput. Eng.*, vol. 8, no. 6, p. 4382, 2018, doi: 10.11591/ijece.v8i6.pp4382-4390.

[7] S. Mingprasert and R. Masuchun, “Applied Artificial Bee colony algorithm for multiple capacitated vehicle routing problem: Case study of the plastic packaging industry,” *ECTI-CON 2017 - 2017 14th Int. Conf. Electr. Eng. Comput. Telecommun. Inf. Technol.*, pp. 270–273, 2017, doi: 10.1109/ECTICon.2017.8096225.

[8] L. Cassettari, M. Demartini, R. Mosca, R. Revetria, and F. Tonelli, “A multi-stage algorithm for a capacitated vehicle routing problem with time constraints,” *Algorithms*, vol. 11, no. 5, 2018, doi: 10.3390/a11050069.

[9] S. E. Comert, H. R. Yazgan, S. Kır, and F. Yener, “A cluster first-route second approach for a capacitated vehicle routing problem: A case study,” *Int. J. Procure. Manag.*, vol. 11, no. 4, pp. 399–419, 2018, doi: 10.1504/IJPM.2018.092766.

[10] J. Clarke, V. Gascon, and J. A. Ferland, “The Case of Medication Delivery and Supervision in the DR Congo,” *IEEE Trans. Eng. Manag.*, vol. 64, no. 3, pp. 327–336, 2017.

[11] I. Sbai, S. Krichen, and O. Limam, *Two meta-heuristics for solving the capacitated vehicle*

routing problem: the case of the Tunisian Post Office, no. 0123456789. Springer Berlin Heidelberg, 2020.

[12] V. R. B. Kurniawan and F. H. Puspitasari, "A mathematical model for delivery zone groups based on courier assignment optimization: A case study in a logistics service provider," *Spektrum Ind.*, vol. 18, no. 2, pp. 183–193, 2020, doi: https://doi.org/10.12928/si.v18i2.17824.

IOP Conf. Series: Materials Science and Engineering 1071 (2021) 012006 doi:10.1088/1757-899X/1071/1/012006

Heating Treatment of Air in Combustion Chamber For The Use of Mixture Ethanol and Gasoline Fuel

A A Ikhsani[1*], M N Kustanto[2] and B A Fachri[3]
[1]Postgraduate Student of Mechanical Engineering, University of Jember, Jember, Indonesia
[2]Department of Mechanical Engineering, University of Jember, Jember, Indonesia
[3]Department of Chemical Engineering, University of Jember, Jember, Indonesia

E-mail: akbaranugrah902@gmail.com

Abstract. This research was conducted to improve the engine performance with ethanol and gasoline mixture, because from fuel mixture with high concentrate causes value of flash point and heat evaporation fuels becomes higher. To overcome these, the air that entrance in combustion chamber heated up to 26°C, 30°C, 40°C and 50°C, the fuel used are E0, E30 and E50 on 2000 to 8000 rotation. The result of research shows E0 fuel highest torque on 30°C with a value 8,98 Nm had 0,22% increase, E30 fuel highest torque on 50°C with a value 9,5 Nm had 1,05% increase and E50 fuel highest torque on 50°C with a value 7,77 Nm had 0,64% increase. On E0 power is 5,96 kW on 30°C had 0,33% increase, the highest power of E30 is 6,29 kW on 50°C had 1,9% increase and the highest power of E50 on 50°C with a value 4,49 kW had 0,89% increase. For the highest consumption is 2000 rotation, for E0 fuel on 50°C with a value 1,43Kg / Hp.Hour, E30 with a value 1,48 Kg/Hp.Hour on 26°C and E50 with a value 2,03 Kg/ Hp.Hour on 26°C, this result shows that air heating treatment can improve the engine performance.

1. Introduction

The application of the use ethanol fuel as a fossil substitute on vehicle becomes one of main steps to reducing pressure on vehicles in using gasoline as main fue[1], but the use of ethanol and gasoline mixture with high concentration can causing problems, that is can reduce the vehicle performance.

The constrains on vehicle if using ethanol fuel as a main fuel is the decreasing of engine performance, that causes by the difference of ethanol and gasoline fuel, which are latent heat of ethanol fuel has triple bigger than gasoline fuel [2], while generally vehicle has technology for gasoline fuel, if the vehicle user wants to use ethanol as a main fuel or mixture fuel, so it must change the technology on vehicles.

The current technology that has been done to adjust the use of pure ethanol fuel or the mixture of gasoline and ethanol is changing the ignition angle and compression ratio on vehicle [3], on this case changing the ignition angle aim to adjust gasoline and ethanol fuel flash point, because the flash point value of gasoline and ethanol fuel were different, it is expected that by changing the ignition angle of ethanol fuel can burned perfectly and also can increase the vehicle performance [4][5].

In the process of changing ignition angle it needed tools that can change the ignition angle on vehicle, it need to consider in the use of ethanol fuel because the ignition angle changing process must be done by using that tools. While changing the compression ratio to be higher aim to maximize energy from ethanol fuel because it needs the higher compression than gasoline fuel compression ratio, the changes of compression ratio must be change the vehicle specification significantly because it has to reduce the thickness of head cylinder [2][6].

In the treatment process to maximize vehicle performance by using of ethanol fuel there is one treatment that can be used without changing the component or system on vehicle that is heating the intake air in the combustion chamber, air intake heating aim to give energy on the air that will come into the combustion chamber [7] so that the intake air in combustion chamber has hot temperature because ethanol fuel has lower temperature than gasoline, the way of air intake works almost same with turbo charge process, but on this research the heating uses a nickelin wire that wrapped around air intake pipe, so when the air mixing with ethanol in the combustion chamber make the ethanol fuel temperature increasing by the mixture of hot air in the combustion chamber [3].

In this research the intake air heated in an air filter box before entering the combustion chamber, air heated using heater from nickelin wire without changing the engine component, in this research only added the heater on air intake pipe in the air filter. In this case it expected can increase the engine performance by using the mixture of ethanol and gasoline fuel with high concentration without changing the engine component.

2. Experimental Method

The heating air using nickelin wire that controled by a digital thermostat that used to controled the intake air temperature in the combustion chamber. The experiment was conducted on experimental with air intake temperature 26°C (standart), 30°C, 40°C and 50°C for every air temperature treatment, the fuel used are E0, E30 dan E50, and for crankshaft rotation that used 2000 Rpm – 8000 Rpm by the multiplier 1000 for every found increase.

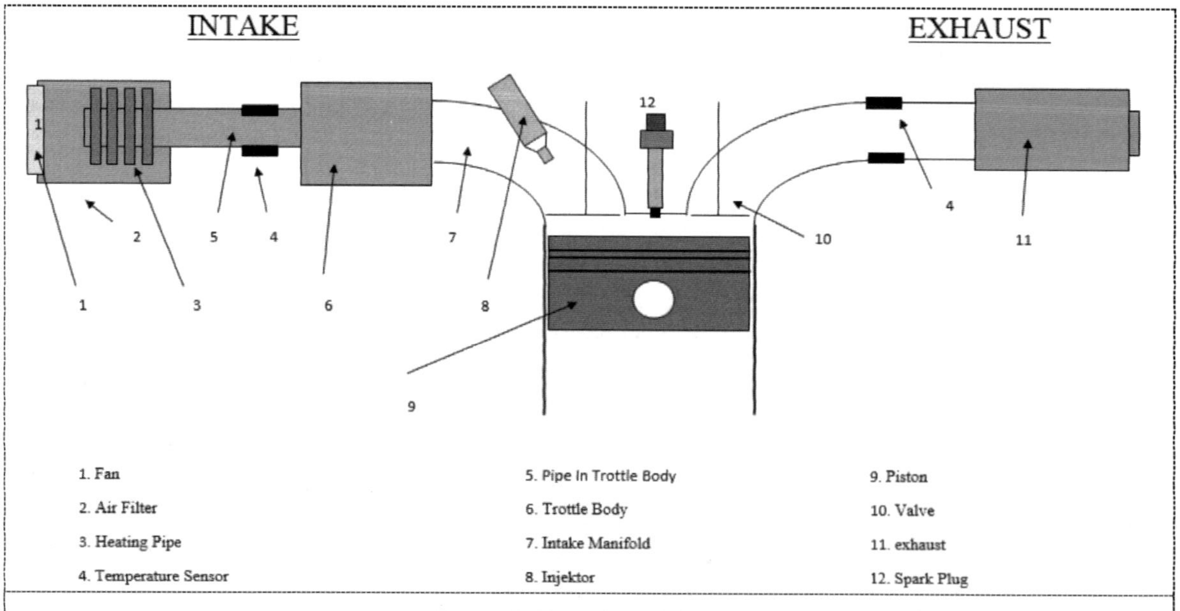

Figure 1. Experimental setup

Table 1. Fuel properties

Properties	**Ethanol**	**Gasoline**
Molecular formula	C_2H_5OH	C_7H_{17}
Molecular weight	46 kg/kmole	100-110 kg/kmole
Density	785 kg/m^3	720-780 kg/m^3
Latent heat of vaporization	904 kJ/kg	350kJ/kg
Calorific value	26800 kJ/kg	43850 kJ/kg
Stoichiometric air/fuel ratio	9.0	14.6
Octane number	106-110	91-96

Table 2. Percentage ethanol and gasoline fuel

Fuel	Percentage
E0	Gasoline 100% + Ethanol 0%
E30	Gasoline 70% + Ethanol 30%
E50	Gasoline 50% + Ethanol 50%

The research done by using gasoline engine by tested the engine perfomance and specific fuel consumption, in this research the engine performance testing using prony brake tools.

Tabel 3. Engine specification

Type Engine	: 4 stroke, SOHC With air cooling, eSP
Stroke Volume	: 108,2 cc
Fuel System	: Injeksi (PGM-FI)
Diameter x Stroke	: 50 x 55,1 mm
Type Tranmision	: Otomatic, V-Matic
Compression Ratio	: 9,5 : 1
Maximum Power	: 6,38 KW (8,68 PS)/7.500 rpm
Maximum Torque	: 9,01 N.m (0,92 kgf.m)/6.500 rpm

3. Result

By using experiment setup on Figure 1 on fuel E0, E30 dan E50 by heating temperature treatment 26°C, 30°C, 40°C dan 50°C get the result a torque, power and specific fuel consumption is presented in the Figure 2, Figure 3 dan Figure 4.

3.1. Torque

From Figure 2 shows the highest torque that produced by E0 fuel is 8,98 Nm on temperature 30°C with increased torque 0,22%, for E30 the highest torque 9,5 Nm on temperature 50°C with increased torque 1,05% and E50 fuel the highest torque 7,77 Nm on temperature 50°C with increased torque

0,64%. The change of air temperature can changed engine performance, in this research the change of air intake combustion chamber can increased torque from engine, especially for the use E30 and E50 fuel, because for ethanol fuel has high flash point is 12°C while gasoline -42°C so that the air heating can increase the air temperature and fuel in the combustion chamber so that the mixture of ethanol and gasoline become a homogeneous can burning perfectly[8].

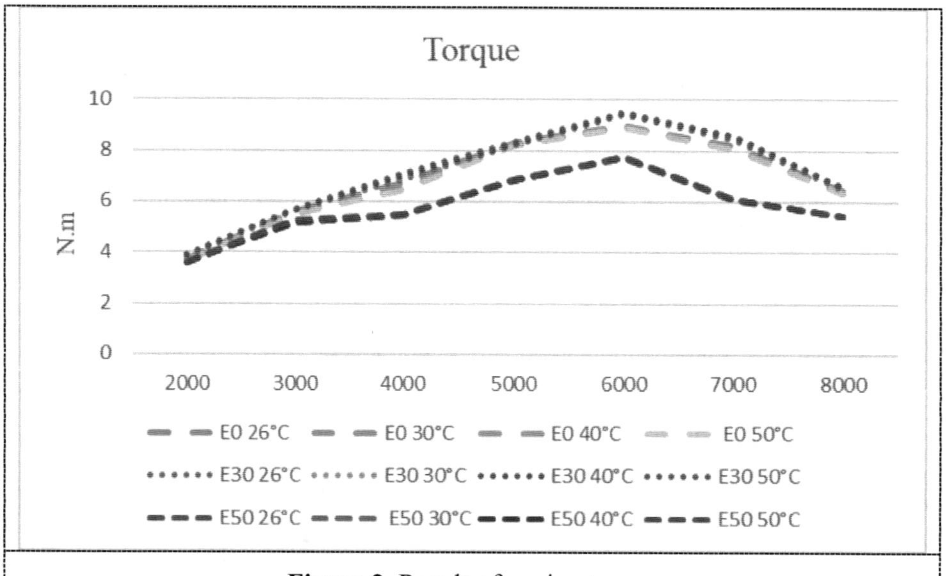

Figure 2. Result of engine torque

3.2. Power

From Figure 3 shows the highest power from fuels E0, E30 and E50 on 7000 rotation that each has value 5,96 KW on temperature 30°C the power has increased 0,33% on E0 , for E30 6,29 KW on temperature 50°C has power increased 1,9% and on E50 with velue 4,49 KW on temperature 50°C the power has increased 0,89%. The result is effected by calorific value, flash point and air temperature treatment which can effected the power output that produced by engine[9][5][10].

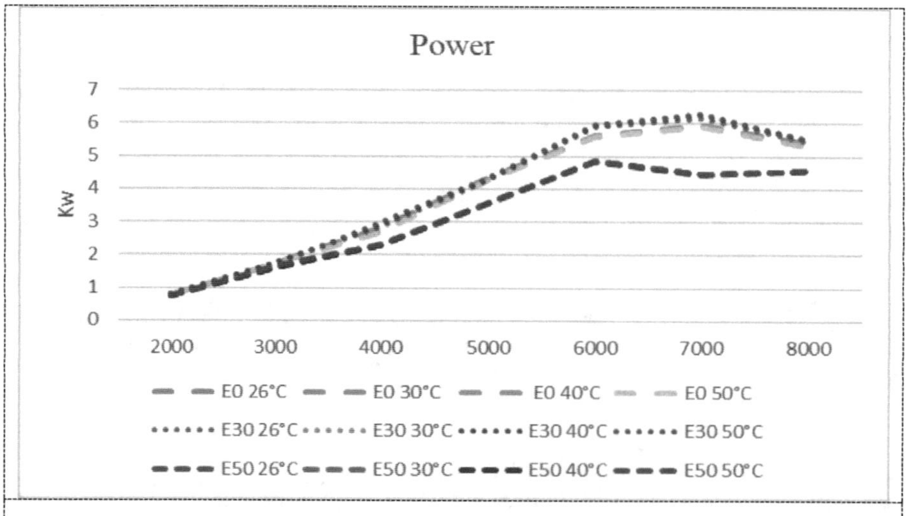

Figure 3. Result of engine power

3.3. Specific fuel Consumption

From Figure 4 to measuring the fuel consumption the test done by calculate time that needed to spend fuel E0, E30 and E50 for each fuel 50 ml, this case has done to know the fuel consumption level from each fuel used . On 2000 rotation the highest consumption E0 fuel on temperature 50°C with value 1,43 Kg/Hp.Hour, for E30 fuel the highest consumption on 2000 rotation on temperature treatment 26°C with value 1,48 Kg/Hp.Hour dan E50 fuel the highest fuel consumption on temperature 26°C with value fuel consumption 2,03 Kg/Hp.hour. The air heating also can decrease fuel consumption it proven on temperature 40°C on E30 fuel that become a lowest consumption from other fuels, the air heating makes air temperature in combustion chamber increased so the air and fuel in the combustion chamber become a homogeneous, if the use mixture of ethanol and gasoline with high concentration can increased the consumption of fuel use on engine, this case also happen on previous research that the consentration mixture of ethanol fuel and gasoline can effected the fuel consumption[6][11][12].

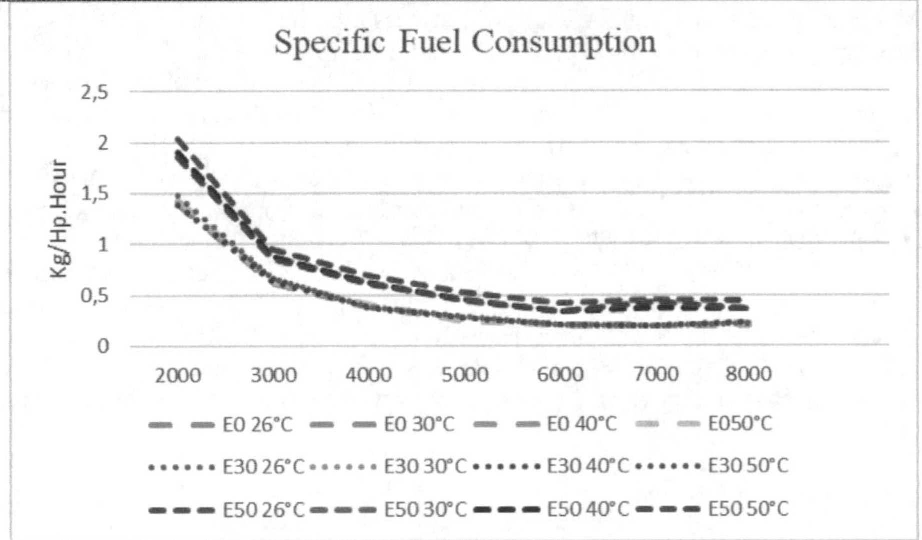

Figure 4. Specific fuel consumption engine

4. Conclusion

In this research found that the intake air combustion chamber treatment can change the engine performance become more increased on each the used of fuel E0, E30 and E50 heating treatment become a solutionto increase the engine performance with a high fuel mixture concentration. Its proven on this research the maximum increased torque is 1,05% on temperature 50°C with E30 fuel, increase in maximum power of 1,9% on temperature 50°C by using E30 fuel and the lowest specific fuel consumption on E30 on temperature 40°C with value 1,39 Kg/Hp.Hour.

References

[1] M. O. El-Faroug, F. Yan, M. Luo, and R. F. Turkson, "Spark ignition engine combustion, performance and emission products from hydrous ethanol and its blends with gasoline," Energies, vol. 9, no. 12, 2016.

[2] S. Phuangwongtrakul, W. Wechsatol, T. Sethaput, K. Suktang, and S. Wongwises, "Experimental study on sparking ignition engine performance for optimal mixing ratio of ethanol-gasoline blended fuels," Appl. Therm. Eng., vol. 100, pp. 869–879, 2016.

[3] P. Sementa, B. Maria Vaglieco, and F. Catapano, "Thermodynamic and optical

characterizations of a high performance GDI engine operating in homogeneous and stratified charge mixture conditions fueled with gasoline and bio-ethanol," Fuel, vol. 96, pp. 204–219, 2012.

[4] M. Paloboran, U. N. Makassar, and B. Sudarmanta, "Erratum to Performances and Emissions Characteristics of Three Main Types Composition of Gasoline-Ethanol Blended in Spark Ignition Engines (International Review of Mechanical Engineering (I.RE.M.E.), 10, 7, (552-559), 10.1016/j.energy.2015.07.004)," Int. Rev. Mech. Eng., vol. 11, no. 1, 2017.

[5] C. Cinar, A. Uyumaz, H. Solmaz, F. Sahin, S. Polat, and E. Yilmaz, "Effects of intake air temperature on combustion, performance and emission characteristics of a HCCI engine fueled with the blends of 20% n-heptane and 80% isooctane fuels," Fuel Process. Technol., vol. 130, no. C, pp. 275–281, 2015.

[6] A. K. Thakur, A. K. Kaviti, R. Mehra, and K. K. S. Mer, "Progress in performance analysis of ethanol-gasoline blends on SI engine," Renew. Sustain. Energy Rev., vol. 69, no. December 2015, pp. 324–340, 2017.

[7] Y. Huang and G. Hong, "Investigation of the effect of heated ethanol fuel on combustion and emissions of an ethanol direct injection plus gasoline port injection (EDI + GPI) engine," Energy Convers. Manag., vol. 123, pp. 338–347, 2016.

[8] A. K. Agarwal, "Experimental investigation on intake air temperature and air-fuel ratio dependence of random and deterministic cyclic variability in a homogeneous charge compression ignition engine," SAE 2011 World Congr. Exhib., no. Dici, 2011.

[9] J. Serras-Pereira, P. G. Aleiferis, D. Richardson, and S. Wallace, "Characteristics of ethanol, butanol, iso-octane and gasoline sprays and combustion from a multi-hole injector in a DISI engine," SAE Int. J. Fuels Lubr., vol. 1, no. 1, pp. 893–909, 2009.

[10] X. Deng, Z. Chen, X. Wang, H. Zhen, and R. Xie, "Exhaust noise, performance and emission characteristics of spark ignition engine fuelled with pure gasoline and hydrous ethanol gasoline blends," Case Stud. Therm. Eng., vol. 12, pp. 55–63, 2018.

[11] A. K. Thakur, A. K. Kaviti, R. Mehra, and K. K. S. Mer, "Performance analysis of ethanol–gasoline blends on a spark ignition engine: a review," Biofuels, vol. 8, no. 1, pp. 91–112, 2017.

[12] P. Chansauria and R. K. Mandloi, "Effects of Ethanol Blends on Performance of Spark Ignition Engine-A Review," Mater. Today Proc., vol. 5, no. 2, pp. 4066–4077, 2018.

The Sidewalk Structuring Strategy As Iconic Pedestrian Case Study: Pedestrian's Budi Luhur University - Jakarta

Dody Kurniawan[1*] and Alvin Yulianto[2]
[1]Lecturer in Prodi Arsitektur, Fakultas Teknik, Universitas Budi Luhur, Jakarta, Indonesia 12260
[2]Student in Prodi Arsitektur, Fakultas Teknik, Universitas Budi Luhur, Jakarta, Indonesia 12260

E-mail: dody.kurniawan@budiluhur.ac.id

Abstract. Over time, sidewalks do not only function as a space for circulation/ movement of pedestrians. The sidewalk is expected to be able to improve the quality of the space where the sidewalk is located and can be an identity/ icon for the surrounding environment. The existence of the sidewalks at the Budi Luhur University Campus has not been properly arranged so that it cannot yet be an identity/ iconic for Budi Luhur University. From the formulation of the problem, research will be conducted with descriptive methods. Descriptive method is used to make a systematic, factual and accurate description, facts, properties and relationships between the phenomena investigated. This depiction is obtained through field observations, interviews and data from relevant agencies. The method of approach taken to achieve the stated goals is Identification of Design Norms; Identification of Comparative Study and Identification of Potential and Problems. The results of this study in the form of the concept of pavement arrangement as an iconic pedestrian.

1. Introduction

Over time, the role of sidewalks today is not only a facility for the movement of pedestrians. The function of sidewalks can be developed as educational facilities or facilities[1] and become a transit center for an area[2]. This condition is in line with the dynamic development of the city and the availability of land that is increasingly narrow. The more city residents who use the sidewalk, the better the city services provided to its citizens. A good city is a city that is fun for pedestrians to do walking activities, enjoy the face of the city and just sit waiting for public transportation. Sidewalks must also become the identity of the city and the environmental identity of an area.

As in general sidewalks in cities, sidewalks in the campus environment are expected to become regional identities. The presence of a campus is the same as the existence of a city, where the campus is a small part of the city environment. So that the identity of a campus can become a city identity. By making sidewalks and drainage as important entities of a road, road corridors will be able to become corridors that support the formation of urban aesthetics[3].

The existing condition of the Budi Luhur University (UBL) campus sidewalks still functions as a facility for moving pedestrians from one building to another. Even the sidewalks at the main field location (between unit 6 and unit 7) are cut off by the vehicle lane. The middle field is strategic because it is right in the middle of the campus site and is the main location for all campus activities, both formal and informal. Based on the above conditions, this study aims to improve the function of the sidewalk in the main field as well as to become the identity of the Budi Luhur University campus area.

2. Problems Of The Study

2.1. Identification
The sidewalks of the Budi Luhur University campus are not well connected and have not become a regional identity.

2.2. Scope Of The Study
In this study the problems discussed are limited to the following:
2.2.1. The sidewalk design that can become a Pedestrian Iconic in terms of comfort and regional identity aspects
2.2.2. The area studied in this study is the middle field area of the Budi Luhur University campus

2.3. Core Of Problems
Based on the research background described above, the existence of sidewalks (pedestrian ways) on the Budi Luhur University campus has the following problems:
2.3.1. There are several locations that have no sidewalks, making it dangerous for pedestrian safety
2.3.2. For sidewalk locations in the central field, the sidewalk design cannot yet become an area's

3. Methodology
This study uses a descriptive research method approach. Descriptive method is carried out to make a sytematical, factual, and accurate description regarding the facts, characteristics and relationships between phenomena investigated[4]. The results of this study are in the form of a redesign of the sidewalk, which functions as a Transit Pedestrian so that it can accommodate the safety and comfort of pedestrians and sidewalks that have become iconic pedestrians. The following are the described methods of approach taken to achieve the predetermined goals:

3.1. Identification Of Design Norms
This part aims to get formulate criteria, indicators, and elements that must be considered in designing sidewalks that accommodate accommodate the safety and comfort of pedestrians and sidewalks that have become iconic pedestrians.

3.2. Identification Of The Comparative Study
This part aims to get a comparison of the criteria, indicators, and elements of pedestrian design that are already there and are considered successful.

3.3. Identification Of Potential And Problems
This identification aims to get determine the condition of the existing research area, both its potentials and problem

3.4. The Stages Of Study
The stages of the research "Sidewalk Structuring Strategy as Transit Pedestrian" are as follows:
3.4.1. The initial stage of research is to conduct a literature study on sidewalks, sidewalk comfort standards and the concept of iconic pedestrians.
3.4.2. The next step is to analyze the iconic aspects that will be used in the research.
3.4.3. The results of the analysis show that the iconic aspects will be used in the Budi Luhur University Campus

3.5. The Factor Of Measure

In the systematic research process, research variables are used to prove hypotheses and to answer existing problems. In the research "Sidewalk Structuring Strategy as a Iconic Pedestrian" used the independent variable and the dependent variable. The independent variable is a variable that has an influence on other variables. The independent variable in this study is the form of sidewalks on Universitas Budi Luhur. The dependent variable in this study is the safety and comfort aspects and iconic aspects.

3.6. Data Collection and The Technique Of Analysis

3.6.1. Determine the Research Area

The first step in this study is to determine the study area. The researcher determines the area to be carried out by the study is the the central field area of the Budi Luhur University campus unit. Primary data are obtained directly from the data source. Primary data is also referred to as original data or new data that has the nature of up to date. To get primary data, researchers must collect it directly. Techniques that researchers can use to collect primary data by visiting, observing directly to the location of study analysis.

3.6.2. Identification of Design Norms

Secondary data obtained through Books/ journals on sidewalk arrangements that accommodate pedestrian safety and comfort and iconic aspects. The output will be in the form of criteria, indicators and design elements considered in the design of the sidewalk (pedestrian ways). Sidewalk planning standards (Departemen Pekerjaan Umum), and urban architectural design standards. The output that will be produced in the form of minimum dimensions that must be met in the arrangement and design of the pavement components.

3.6.3. Identification of Comparative Study

The comparative study cases conducted was a visit to sidewalk area of Losari Makassar and Martapura's River sidewalk. The selection of the case study site is based on the assessment that the two sidewalks are considered successful in applying the sidewalk principle that accommodates pedestrian safety and comfort and iconic aspect.

4. Discussion

4.1. The Location of Study

The study location is in the middle field of the Budi Luhur University campus. The selection of this location is based on several criteria:

4.1.1. It is the central point of the Budi Luhur University campus site plan

4.1.2. To become the focal point of the Budi Luhur University campus activities

IOP Conf. Series: Materials Science and Engineering 1071 (2021) 012007 doi:10.1088/1757-899X/1071/1/012007

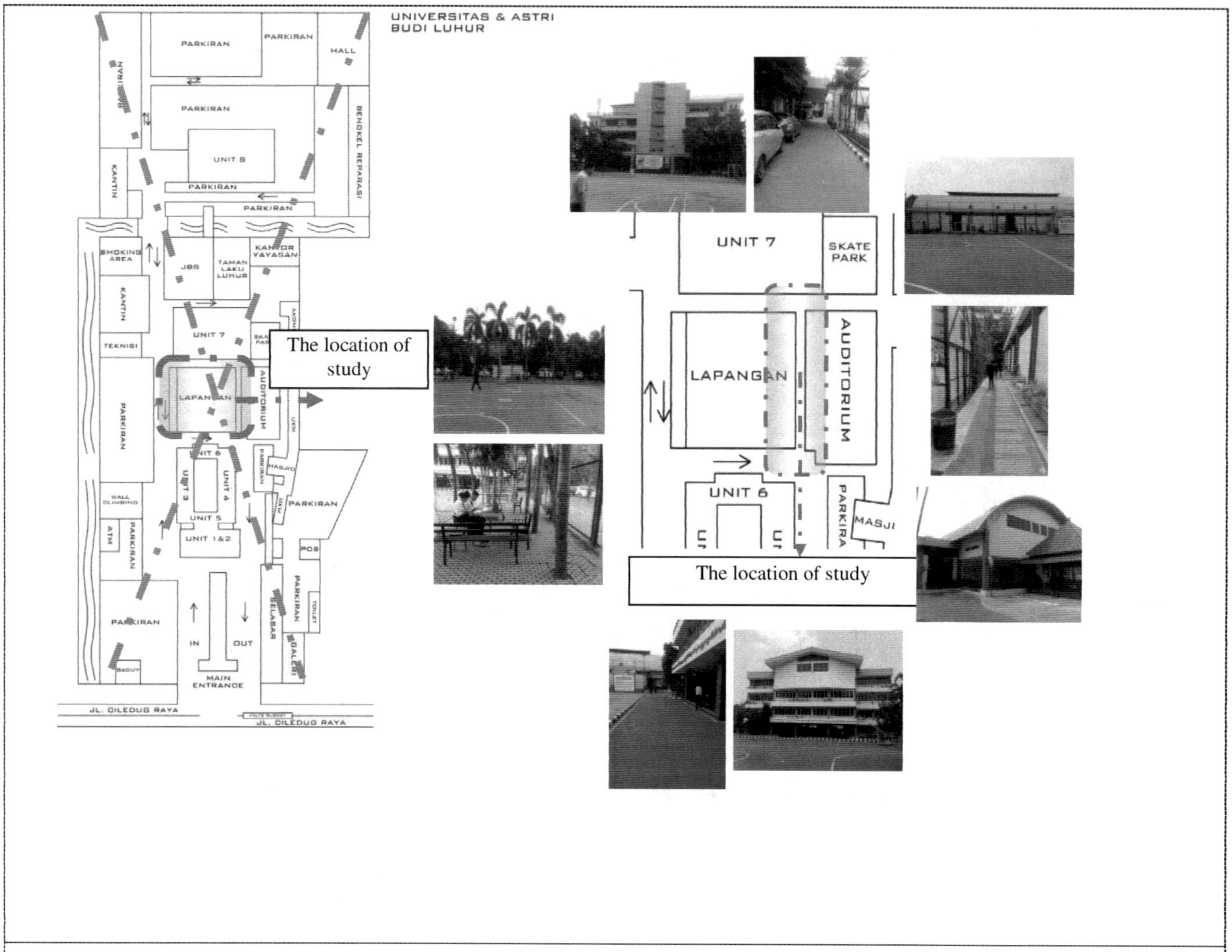

Figure 1. The location of study (left); Existing conditions of around study area (right)

The study location is a connecting road between Unit 6 and Unit 7 on the Budi Luhur University campus. This road is bordered by the field on the west side and the auditorium on the east side. On the north side, there is the unit 7 campus building and the south side is the unit 6 campus building. This road condition is not supported by pedestrian facilities. Pedestrians must pass through motorized vehicles if they want to move from unit 6 to unit 7 of the Budi Luhur University campus building.

The auditorium building (east side) is used for formal meeting activities. This building is commonly used for seminars, wedding receptions and other activities that require privacy. The field (west side) is used for sports activities, competitions, and activities that require open space. Unit 7 (north side) and Unit 6 (south side) are used as the learning building.

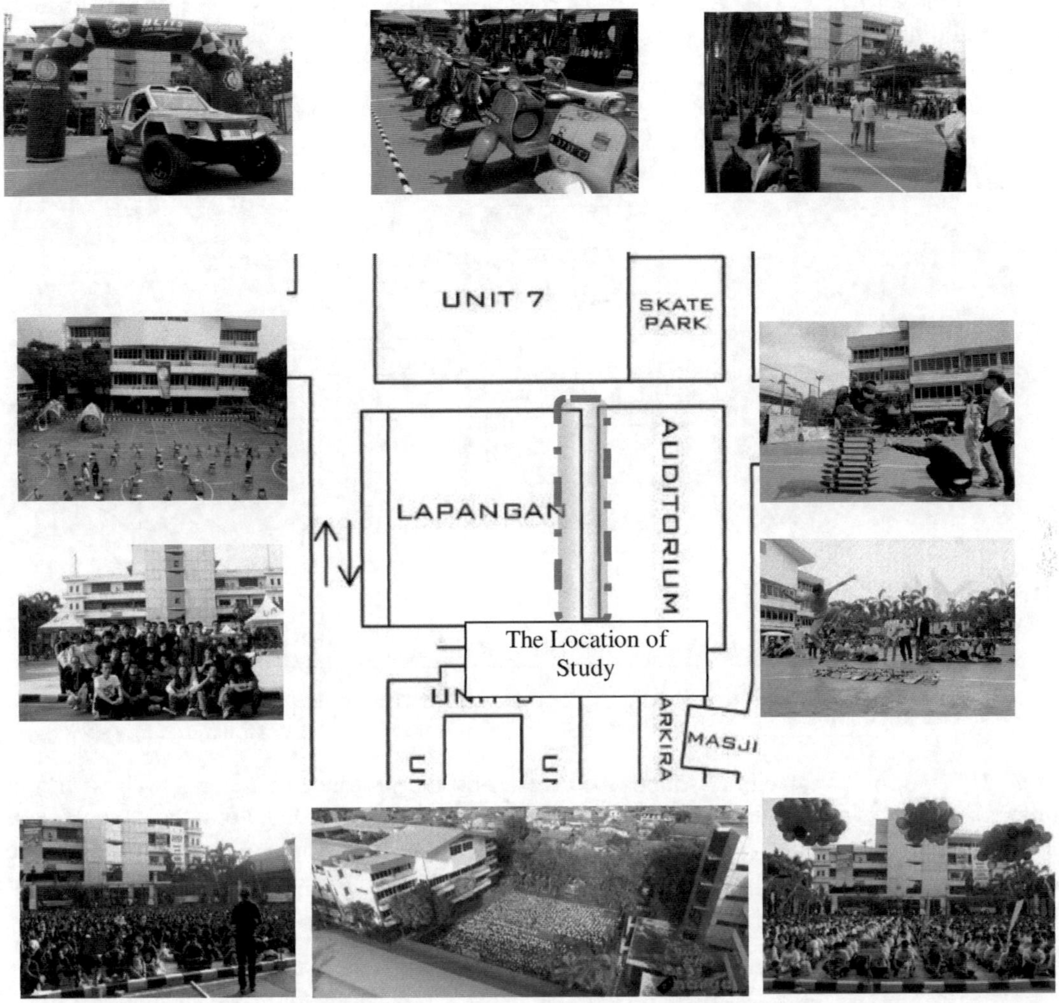

Figure 2. Activities in the main field

4.2. The Result of The Study

Based on the results of the analysis of the research aspects in the form of shapes and colors, the following results were obtained:

4.2.1. Shape

Based on strategic location points, it is necessary to have a form of building mass that can become a focal point. This research takes the object of a rotating rectangular shape. The existence of this bridge is not only a focal point, but also a means of unification among other building objects around the Budi Luhur University field. This bridge is not to function itself as an element that is more important than others, but to strengthen clarity of spatial functions, direct focus, and sharpen existing spatial concepts.

This bridge (sky bridge) is located on the east side of the Auditorium building, which connects Unit 6 (2nd floor) building on the South side with Unit 7 (2nd floor) on the North side of the field. Prior to this bridge, Unit 6 pedestrians who wanted to go to Unit 7 (or vice versa) had to walk in the hot sun. With this bridge, it is hoped that pedestrians in Unit 6 and Unit 7 can move places without having to get off. Meanwhile, pedestrians on the ground floor still get comfortable walking because this bridge will protect the sun's heat. The selection of the rotated rectangular shape is based on the concept of the existing building shape around the site.

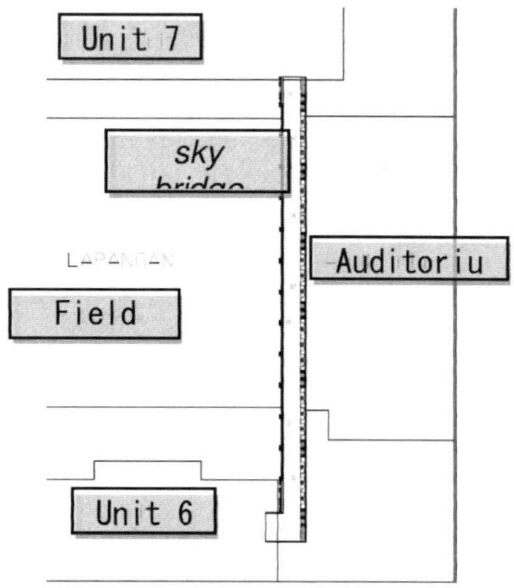

Figure 3. Map of the connecting bridge location

The selection of a rectangular shape is enhanced by the presence of a panel wall with a circular motif. To give a strong and firm impression, a list of columns was added along the connecting bridge. For the safety aspect, the bridge is equipped with railings on both sides of the bridge. The choice of glass material gives a broad impression of the interior (interior) of the connecting bridge. Pedestrians can enjoy the atmosphere outside the bridge or the interesting interior atmosphere.

Figure 4. Map of the connecting bridge location

The connecting bridge is equipped with interior facilities that support learning activities. The wooden bench equipped with flower pots gives a natural impression and a comfortable atmosphere for the learning process. Wooden benches are placed along the bridge without disturbing pedestrian circulation. The inside walls are covered and decorated with graffiti themed with the messages of the founder of the Budi Luhur Foundation. In order not to look monotonous, the glass walls are adorned with circular patterned panels. The upper part of the bridge is equipped with lighting so that the learning atmosphere becomes more comfortable

Figure 5. The interior of a glass wall (left). The interior of the bridge (center). The section of bridge (right)

4.2.2. Color

Color is another aspect analyzed in this study. At the bottom of the bridge, a batik motif is given with the inscription UNIVERSITAS BUDI LUHUR. This point is expected to become a "Selfi Spot" that can support activities around the campus field. The combination of gold and glass gives a luxurious impression and can characterize the connecting bridge icon. The bridge column list is given a silver color so that it gives a strong and firm impression.

5. Conclusions

After analyzing the aspects that affect the iconic pedestrian arrangement design:
The circulation aspect is obtained from the existence of a sky bridge. Pedestrians who are in building units can immediately move and walk without having to leave the building.

Aspects of pedestrian comfort and safety are obtained from the semi-closed sky bridge. Pedestrians can walk and move places without having to get hot and in the rain and avoid cross circulation with motorized vehicles

The visual aspect is obtained by using a rotated rectangular shape on the sky bridge and color finishing both on the laying bridge and the walls around the existing field sidewalk.

References

[1] Dody Kurniawan 2020, Pedestrian Transparent as One of Education Means and Disable's Friendly: Case Study - Pedestrian's Universitas Budi Luhur (SCITEPRESS); 87-92

[2] Dody Kurniawan 2020, The Sidewalk Transmitting Strategy As A Transit Pedestrian (Case Study: Jalan Manggarai Utara 1, Jakarta-Indonesia) (IOP) vol 436; 1-9

[3] Gustiaji, L., Herianto, D., Hadi, Y. M. (2016), Studi Optimalisasi Perparkiran Dan Pedestrian Di Fakultas Teknik Jurusan Teknik Mesin, Teknik Elektro, Teknik Kimia Dan Teknik Geofisika Universitas Lampung. Jurnal Rekayasa Sipil Dan Desain, 4:1, 33-42

[4] Yunianto, Prasetiyo. The Iconic Stilation Of Malioboro Street Furniture. Jurnal SULUH, 103-118

[5] Safarina, T Irma. (2014), Kontekstual Ikonik Pada Pengembangan Kampus Institut Teknologi Nasional 2030. Jurnal Reka Karsa, 3:2, 1-10

[6] Saraswaty, R. (2017). Kenyamanan Pejalan Kaki Terhadap Pemanfaatan Trotoar Di Jalan Brigjen Katamso Medan. Jurnal Education Building, 3:1, 9-14

[7] Nazir 2005, Metode Penelitian Ghalia Indonesia

IOP Conf. Series: Materials Science and Engineering 1071 (2021) 012008 doi:10.1088/1757-899X/1071/1/012008

Risk Allocation Model For Cisumdawu Toll Road Projects

D Setiawan[1]*, R Milyardi[1], T L Ing[1], C Rizkiana[1]
[1]Department of Civil Engineering, Faculty of Engineering, Universitas Kristen Maranatha, Bandung, West Java, Indonesia

E-mail: deni.setiawan@eng.maranatha.edu

Abstract. In the Cisumdawu Toll Road project there are some complex problems, including the first tunnel structure in Indonesia that functions as a toll road. This complex construction project is currently experiencing delays which creates risks for project delays. In this study the aim is to identify risks and create an optimal risk allocation model between the project owner and the contractor. The allocation of risk between project owner and contractor in a contract plays an important role for the success of the project. It should be based on a proper assessment of the involved risks and choosing the party best able to manage them. This research was obtained about risks that can be impacted by the project in each phase. The scope of non-excusable factors often is considered a contractor's fault which results in project construction. Based on the results of descriptive test of the 5 main risks include the most important risk in the Cisumdawu Toll Road Construction Phase II (STA 9+750 – STA 27+100) is soil work. The results show that the soil work risk should bear by the contractors by lump sum fixed price contract.

1. Introduction

All construction projects are unique and carry their own risks. Such projects involve a number of parties concerned, starting with the owner, contractor, designer, suppliers, and others. All parties involved in a project inevitably carry certain risks. Risk can be defined as a hazard, a probability of it to occur and the potential of losses and resulting gains [1]. Risk can be defined as a difference of actual and expected results [1]. Risks can be managed, reduced, transferred or accepted, but it cannot be ignored [2]. Risk management process usually consists of four stages: risk identification, risk analysis, selection and monitoring of risk management techniques to the consequences of risk. The implementation phase of construction started after the owner and the contractor sign a construction contract. The purpose of construction contract agreement is to allocate the rights, responsibilities and risks between the parties. Construction business is a business with high risks, such as financial risks, political risk, security risk and risk at the time of execution was to be managed and handled by contractors. From the moment that the decision to begin design is taken until the new facility is in use, the owner is uncertain about the outcome of the project.

There are three uncertainties in construction project: time, cost and quality. These uncertainties are the three types of risk that are present in every project and that affect the owner [3]. Risks are usually considered as uncertain future events, which, if they happen, will cause significant extra cost or delay. Generally, the project owner usually prefers to allocate most of the risk to the contractor. The impact of " must" bear a lot of risks , the contractor will perform a " trade off " to the offer price as a consequence to accept the risks that are beyond the control of the contractor [4].

2. Research Method

In this research, data collection was done as a basis for the theory of submission of the questionnaire. Second, the data will be accumulated. Third, an analytic hierarchy process is carried out to determine the main work that is the highest risk in the Cisumdawu toll project. Research method diagram showed in the Figure 1. Justification one main work for testing in the Analytical Hierarchy Process (AHP) [5] within the stage show in Figure 2.

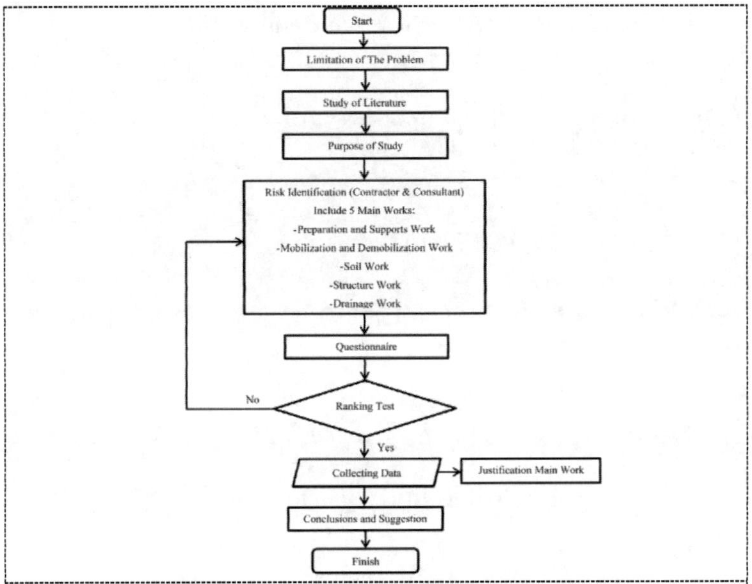

Figure 1. Research Flow Chart Diagram

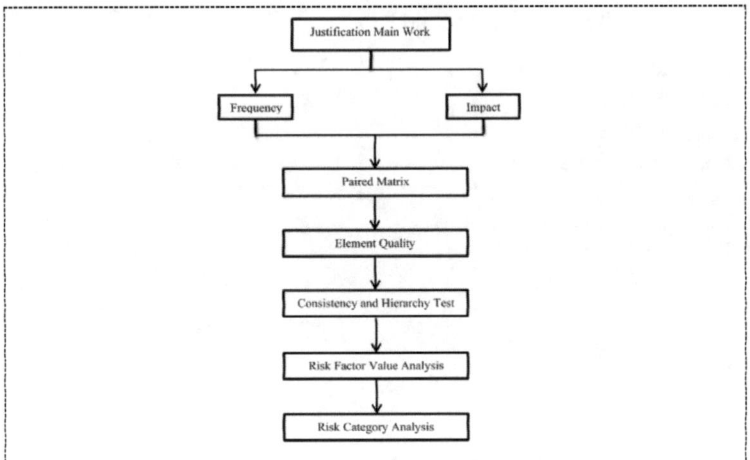

Figure 2. AHP Test Flow Chart Diagram

This research uses a survey method in the form of a questionnaire, where carried out by analyzing the research and final project that has been done in the field of transportation especially in highway project. Then, the next step is to look for factual information about the conditions that indeed occur in the project that is focused on the cisumdawu phase 2 toll road project.

3. Analysis Result

The result that can be classified all risks in each job into a high level that must be considered when the highway project was installed. Cisumdawu toll road phase II project respondents have a different perception; they assume the main work that should be considered is soil work. Furthermore, after interviewing the employees in Cisumdawu Toll Road Project Phase II and they are trying to explain about most common obstacles that occur in soil work. This can be led to differences of opinion due to different places, equipment, and workers due to the time of implementation highway project. Then, the focus of the analytical hierarchy process (AHP) test is the soil work.

3.1. Risk Analysis

After obtaining the average of the impact and frequency risk value, risk factor value is then performed, which can be calculated with the following equation [5]:

$$FR=L+I-(L\times I) \quad (1)$$

FR = Risk Factor, scale 0 -1
L = Risk Probability
I = Risk Impact

Table 1 shown recapitulation value results from aggregate variable/ risk occasion:

Table 1. Risk Factor Value

Sub Indicator C	Frequency Average Value	Impact Average Value	Risk Factor (FR)
Unpredictable of weather.	0,441	0,421	0,676
Land that has not been cleared. Hence, the project obstructed.	0,444	0,504	0,724
Obstacles on mobilization and demobilization for aggregate is not on time.	0,148	0,168	0,291
Productivity of employees	0,180	0,180	0,327
Working methods are less precise.	0,200	0,193	0,354
Lack of policy application in OSH (Occupational Safety and Health) aspects.	0,160	0,160	0,295
Damage residential area during the project construction.	0,215	0,602	0,687
Respiratory problems due to material dust for workers and local residents.	0,180	0,122	0,280
The land that cannot be used after being cleared.	0,179	0,147	0,299
Risk of soil/site existing conditions.	0,122	0,135	0,240

Sub Indicator C	Frequency Average Value	Impact Average Value	Risk Factor (FR)
Obstruction of public transportation access	0,140	0,109	0,234
Limitation of workspace and co-working space construction	0,115	0,122	0,223

A qualitative risk assessment can also help to determine if there are any specific types or categories of risks that would require special attention or any risk events that need to be handled in the near-term. Then, I tried to make the qualitative risk analysis matrix on the Figure 3.

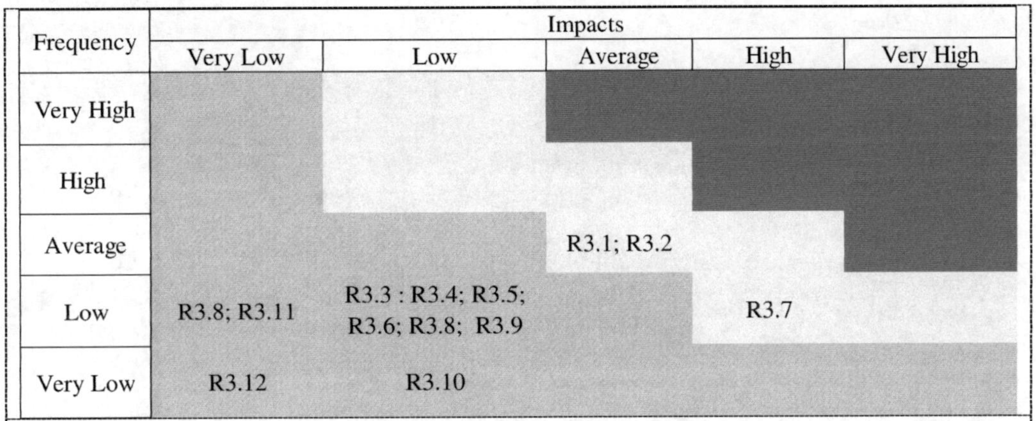

Figure 3. Qualitative Risk Analysis

Risks (R) information:
R3.1 Unpredictable of weather.
R3.2 There is land that has not been cleared. Hence, the project obstructed.
R3.3 Obstacles on mobilization and demobilization for aggregate is not on time.
R3.4 Productivity of employees
R3.5 Working methods are less precise.
R3.6 Lack of policy application in OSH (Occupational Safety and Health) aspects.
R3.7 There is damage residential area during the project construction.
R3.8 Respiratory problems due to material dust for workers and local residents.
R3.9 There is the land that cannot be used after being cleared.
R3.10 Risk of soil/site existing conditions.
R3.11 Obstruction of public transportation access
R3.12 Limitation of work space and co-working space construction

Table 2 shown results of risk category in soil work by AHP method.

Table 2. Results of Risk Category in Soil Work by AHP

Sub Indicator C	Risk Factor (FR)	Risk Level
3.1 Unpredictable of weather.	0,676	Average
3.2 Land that has not been cleared. Hence, the project obstructed.	0,724	High

Sub Indicator C	Risk Factor (FR)	Risk Level
3.3 Obstacles on mobilization and demobilization for aggregate is not on time.	0,291	Low
3.4 Productivity of employees	0,327	Low
3.5 Working methods are less precise.	0,354	Low
3.6 Lack of policy application in OSH (Occupational Safety and Health) aspects.	0,295	Low
3.7 Damage residential area during the project construction.	0,687	Average
3.8 Respiratory problems due to material dust for workers and local residents.	0,280	Low
3.9 The land that cannot be used after being cleared.	0,299	Low
3.10 Risk of soil/site existing conditions.	0,240	Low
3.11 Obstruction of public transportation access	0,234	Low
3.12 Limitation of workspace and co-working space construction	0,223	Low

The results of Risk category in soil work shows that is categorized as high risk is 3.2, and data which categorized as average risk is 3.1; 3.7. Then, for variable that categorized as low risk is 3.3; 3.4; 3.5; 3.6; 3.8; 3.9; 3.10; 3.11; 3.12.

3.2. Risk Allocation

The most important risk in the Cisumdawu Toll Road Construction Phase II (STA 9+750 – STA 27+100) is soil work. The origin of risk factors was then analyzed and discussed based on Indonesia's Risk Allocation Guidelines from PT. Penjaminan Infrastruktur Indonesia, all technical risks should be borne by the contractors. So, soil work risk must bear with contractor party.

4. Conclusion

The result of this study shows that contractors bear important risks that impact to delay in construction phase. All risk factor in soil work allocate to contractor party by lump sum fixed price contract mechanism. The result at this study will contribute to provide the lesson learned to develop a risk management for toll road project in Indonesia.

References

[1] Lam K C, Wang D, Lee P T K and Tsang Y T 2007 Modelling risk allocation decision in construction contracts Int. J. Proj. Manag. 25 485–93
[2] Khazaeni G, Khanzadi M and Afshar A 2012 Optimum risk allocation model for construction contracts: fuzzy TOPSIS approach Can. J. Civ. Eng. 39 789–800
[3] Barnes M 1983 How to allocate risks in construction contracts Int. J. Proj. Manag. 1 24–28
[4] Turskis Z, Gajzler M and Dziadosz A 2012 Reliability, Risk Management, And

Contingency Of Construction Processes And Projects / Statybos Procesų Ir Projektų Patikimumas, Rizikos Ir Atsitiktinumų Valdymas J. Civ. Eng. Manag. 18 290–8
[5] Saaty T L 2004 Decision making — the Analytic Hierarchy and Network Processes (AHP/ANP) J. Syst. Sci. Syst. Eng. 13 1–35

Physical Properties of Liquid Soap using Katuk Leaf Extract (sauropus androgynus (l) merr.) as an Alternative to Natural Surfactants

I L P Mursal[1*],V A Hermana[1] and Farhamzah[1]
[1]Universitas Buana Perjuangan Karawang

*E-mail : iin.lidia@ubpkarawang.ac.id

Abstract. Katuk leaves (Sauropus androgynous (L) Merr.) contain saponins that can potentially act as natural surfactants to replace Sodium Lauryl Sulfate (SLS). This study aims to determine the variation in the concentration of katuk leaf extract (Sauropus androgynous (L) Merr.) On the results of testing the physical properties of liquid soap formulations and to determine the potential of katuk leaf extract as an alternative to natural surfactant substitute (Sauropus androgynous (L) Merr) in the formulation. Liquid soap preparations. The liquid soap formulation of katuk leaf extract with concentrations of 1%, 3%, 5%, 7%, and 9% was replicated 5 times. The physical properties test carried out were organoleptic testing, pH, foam height, specific gravity, and viscosity. Based on the research that has been done, it can be concluded that the variation in the concentration of katuk leaf extract affects the physical properties of liquid soap includes organoleptic, pH, high foam, specific gravity, and viscosity. The liquid soap formulation uses katuk leaf extract has potential as an alternative to natural surfactants instead of surfactants. It can be known that the three formulas, namely F1, F2, and F3 as a whole meet SNI standards.

1. Introduction

Liquid bath soap is a primary requirement in daily activities. Increasing of the human population every year results in the use of bath soap also increasing. Soap is generally defined as the alkaline salt of long-chain fatty acids. When the fat or oil is saponified, sodium or potassium salts are formed from long-chain fatty acids called soap. Bath soap is a preparation used by the community to clean the skin. Soap is produced from two main ingredients, namely alkalis and triglycerides [1]. Generally, the bath soap used today uses synthetic chemicals as surfactant active ingredients [2].

The surfactant that is usually used in liquid bath soap formulation is Sodium Laury Sulfate (SLS). Sodium Lauryl Sulfate (SLS) is a widely used anionic surfactant in various nonparenteral pharmaceutical and cosmetic formulations [3]. Sodium Lauryl Sulfate (SLS) is not a carcinogen when applied to the skin but, based on SLS research it can irritate the skin and face when applied for a long time and continuously [4].SLS and ALS can cause severe skin irritation and both of these substances can be easily absorbed into the body. Once absorbed, these substance deposits will be present in the brain, heart, lungs, and liver which will become a long-term health problem. SLS and ALS also have the potential to cause cataracts and interfere with eye health. Synthetic active ingredients can cause negative effects for humans with sensitive skin, which can cause irritation. A solution of just 2% SLS can increase skin thickness, cause irritation, inflammation [5] and increase other forms of immune activity in the skin [6]. SLS can cause an increase in enzyme levels in the skin, leading to redness and swelling [7]. It can also lead to dryness, roughness and even flaking of the skin.

Therefore, it is necessary to research to find alternatives to natural surfactants derived from natural ingredients. The production of natural surfactants aims to minimize the negative impact on the body caused by SLS. Natural ingredients that can potentially be used as natural surfactants are those containing saponins. Saponin is a glycoside that has an aglycone in the form of sapogenin. Saponins can reduce the surface tension of the water so that it will result in the formation of foam on the surface of the water after being shaken. This property has in common with surfactants. The decrease in surface tension is caused by the presence of soap compounds that can damage the hydrogen bonds in water. This soap compound has two parts that are not the same polarity [8]. In this research, research was conducted to see the physical properties of liquid bath soap from katuk leaf extract. The use of katuk leaves because they contain saponin compounds which can be potential as natural surfactants to replace Sodium Lauryl Sulfate (SLS).

2. Material and Method

2.1. Materials

Katuk leaves, Alcohol 70%, Olive oil, Sodium Lauryl Sulfate (SLS), Cocamid DEA,Na-CMC,NaCl, Na-Benzoic Stearic Acid Katuk leaves, Alcohol 70%, Olive oil, Sodium Lauryl Sulfate (SLS), Cocamid DEA,Na-CMC,NaCl, Na-Benzoic Stearic Acid, KOH 20%, Vitamin E, and Aquadest.

2.2. Katuk leaf 70% ethanol extract preparation

500 grams of katuk leaf powder was extracted using 5 liters of 70% ethanol solvent by the maceration method. Katuk leaves are put in a brown bottle then added 75 parts of 70% ethanol (3750 ml), and left to stand for 5 days. After 5 days of repeated shuffling. Macerate is filtered and the residue is squeezed. The residue was added with 25 parts of 70% ethanol (1250 ml), then stirred and obtained 100 parts of the entire juice. The extract obtained is concentrated with a rotary evaporator until a thick extract is obtained. The remaining solvent is evaporated on a water bath until it is solvent-free.

2.3. Preparation of Liquid Soap Formula for katuk leaf extract

Olive oil and stearic acid are heated at 500 °C, adding 20% potassium hydroxide little by little to form a paste. then added Na CMC and NaCl which have been dissolved with a little water until homogeneous. The mixture was transferred to a beaker then added with katuk leaf extract, stirring until homogenous. then add cocamide DEA and sodium benzoate, stirring until homogeneous.
The manufacture of liquid soap for the ethanol extract of katuk leaves was adjusted to each concentration. After that, the quality test of the ethanol extract of katuk leaves liquid soap was carried out by organoleptic test, pH, foam height, specific gravity, and viscosity

Table 1. Variations of katuk leaf extract concentration

Bahan	(%)					
	F_{SLS}	F1	F2	F3	F4	F5
SLS						
Katuk Leaf Extract		1	3	5	7	9
Cocamid DEA	2	2	2	2	2	2
Olive oil	20	20	20	20	20	20
Na-CMC	3	3	3	3	3	3
Na Benzoat	0,1	0,1	0,1	0,1	0,1	0,1
Stearic Acid	2	2	2	2	2	2

NaCl	0,5	0,5	0,5	0,5	0,5	0,5
KOH 20%	3	3	3	3	3	3
Vitamin E	0,05	0,05	0,05	0,05	0,05	0,05
Aquadest	ad 100	ad 100	ad 100	ad 100	ad 100	ad 100

3. Formatting the title, authors and affiliations

In this study, the natural ingredients to be used are Katuk leaves. The use of katuk leaves because they contain saponin compounds which can have the potential as a natural surfactant to replace Sodium Lauryl Sulfate (SLS). Surfactants, or surface-active agents, are amphiphilic molecules. Their heads are polar, or hydrophilic, and their tails hydrophobic. They are soluble in both organic solvent and water. The surfactant reduces the surface tension of water by adsorbing at the liquid-gas interface.

The organoleptic test results on the F sls sample show that the soap is liquid, has a white color, has a strawberry smell. This smell is caused by the addition of strawberry scent to the formulation. the color of F1 and F2 samples are light green while the F3, F4, and F5 formulas are dark green. The green color of the liquid soap indicates the presence of katuk leaf extract in the formulation. Then, the color difference at F3, F4, and F5 is getting darker green due to the increased concentration of katuk leaf extract. from the results of the organoleptic test, all samples have complied with SNI standards.

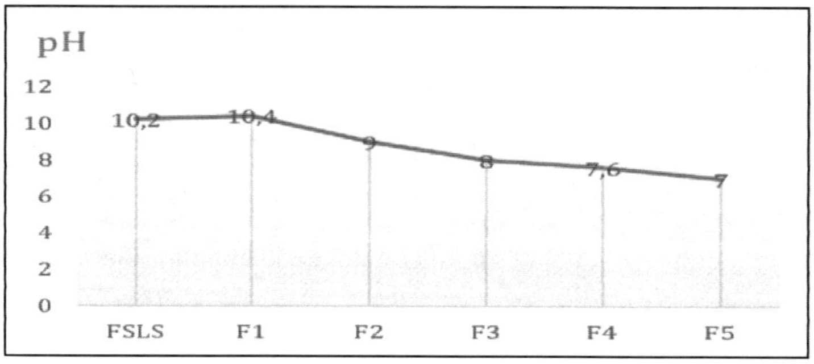

Figure 1. pH at the variation of katuk leaf extract concentration

The comparison of the results of pH testing of liquid bath soap formulations using SLS and using katuk leaf extract can be seen in Figure 1. The pH test is one of the requirements for the quality of liquid soap. This is because liquid soap is in direct contact with the skin and can cause problems if the pH does not match the skin's pH. In general, liquid soap products have a pH that tends to be alkaline. This is due to the basic ingredients of the liquid soap, namely KOH which is used to produce saponification reactions with fats or oils or synthetic detergents that have a pH value above neutral pH.

More the concentration of katuk leaf extract is added, the pH will decrease. pH changes occur because Katuk leaf extract has an acidic pH of 4.5. Formulations that meet the standard of liquid bath soap are formula F1 (pH = 10.4) F2 (pH 9) and F3 (pH 8). Formulations that do not meet the standards within the range of SNI criteria are formula F4 (pH 7,6) ± and formula (F5 pH 7). Changes in pH on the preparation are influenced by the increase in the variation of the katuk leaf extract in liquid bath soap. In addition, a decrease in pH can also be caused by other ingredients that make up soap, namely stearic acid, which is acidic [9].

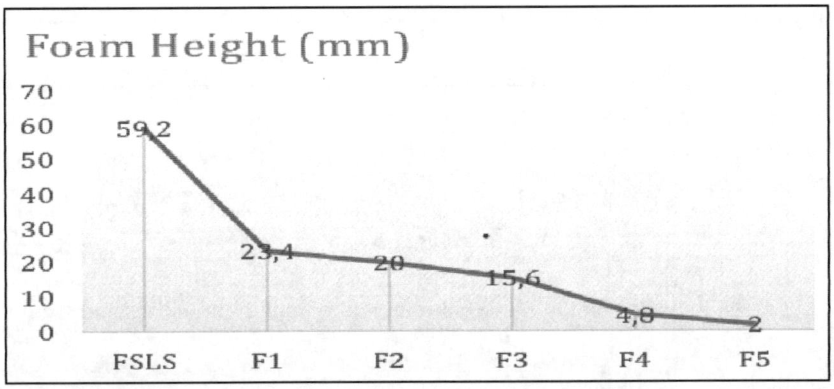

Figure 2. Foam height at the variation of katuk leaf extract concentrations

The increased concentration of liquid soap, the less foam is produced. There is a decrease in the height of the foam because the height of the foam is influenced by pH, so that the lower the pH, the less foam is produced. F1, F2, and F3 have met the SNI standard for foam height (Fig 2). There is a decrease in the height of the foam because the height of the foam is influenced by pH, so that the lower the pH, the less foam is produced. Foam stability is influenced by the concentration and viscosity of the preparation [10]. Based on the results obtained in the F1 formula, the foam height obtained was 23.4 ± 2.07 mm, the F2 formula for the foam height obtained was 20 ± 1.41 mm and the F3 formula for the foam height obtained was 15.6 ± 0.89 mm which met soap standards. liquid. In the formula F4, the foam height obtained is 4.8 ± 1.79 mm and F5 the foam height obtained is 2 ± 0.00 mm, not according to what is stipulated by SNI 06-4085-1996.

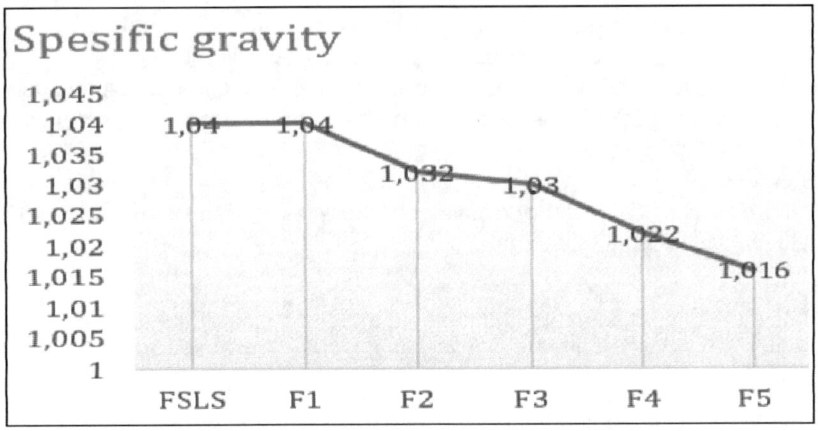

Figure 3. Spesific gravity at the variation of katuk leaf extract concentrations

From the observations, it was obtained that the specific gravity in the F1 formula was 1.04 g / ml, the specific gravity in the F2 formula was 1.032 g / ml, the specific gravity in the F3 formula was 1.03 g / ml, the specific gravity in the F4 formula was 1.022 g / ml and the specific gravity in the F5 formula was 1.016. g / ml. The specific gravity test of each has decreased the specific gravity value but is still within the criteria range. The building weight value of a constituent material and its physical properties. the decrease in specific gravity is caused by the presence of fat or ethanol in the solution.

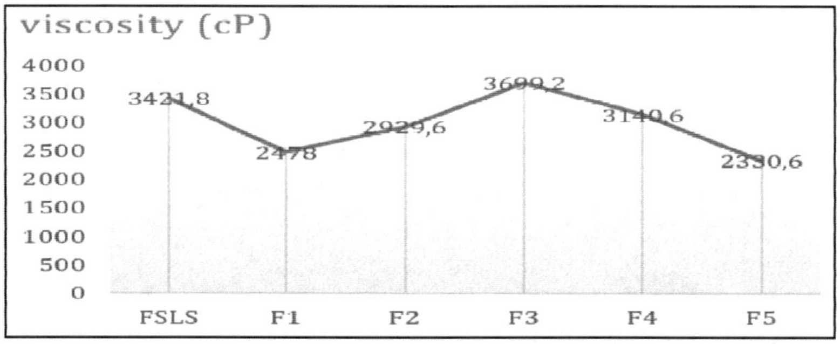

Figure 4. Viscosity at the variation of katuk leaf extract concentrations

Fig 4 shows the results of the study from the comparison of the viscosity value of liquid bath soap formulations using SLS and katuk leaf extract. The viscosity test used the viscometer Lamy Rheology, from the results of observations on the F1 formula the results were 2478 cP, the F2 formula got 2929.6 cP, the F3 formula got 3699.2 cP, the F4 formula got 3140.6 cP and the F5 formula got 2330 results cP. Katuk leaf extracts liquid soap in the formula F1, F2, F3, F4 and F5 have met the viscosity standard of liquid soap. With the balance of the composition between fatty acids and bases, the saponification process runs perfectly, so that the resulting soap product becomes thicker. Also, other factors that cause changes in the viscosity value are the length of stirring in the liquid soap formula and the temperature used in the process of making liquid soap formulas.

4. Conclusion

Based on the research that has been done, it is known that the concentration of katuk leaf extract has an effect on the physical liquid soap including organoleptic, pH, high foam, specific gravity, and viscosity. The liquid bath soap formula for katuk leaf extract that meets all physical properties standards is F1, F2, and F3. Based on these results, it can be seen that liquid soap formulation using katuk leaf extract has the potential as an alternative to natural surfactants to replace SLS.

Acknowledgement

The completion of this study could not have been possible without the support of laboratory team of Buana Perjuangan Karawang University.

References

[1] Anggraini D, Wiwik SR, Masril M. 2012. *Formulasi Sabun Cair dari Ekstrak Batang Nanas (Ananascomosus. L) untuk Mengatasi Jamur Candida albicans. Jurnal Penelitian Farmasi Indonesia* 1(1), September 2012

[2] Nita, A. 2015. *Uji Kandungan Senyawa Fitokimia Kulit Durian sebagai Bahan Aktif Pembuatan Sabun*. Program Studi Teknik Kimia, Universitas Serang Raya

[3] Allen, LV., 2009, *Handbook of Pharmaceutical Excipients, Sixth Edition*, Rowe R.C., Sheskey, P. J., Queen, M. E., (Editor), London, Pharmaceutical Press and American Pharmacists Assosiation, 697-699

[4] Awim, D. A.,2017. *Pengaruh Penambahan Konsentrasi Larutan Surfaktan Sodium Lauryl Sulfate (Sls) Terhadap Tegangan Permukaan Dan Viskositas Oli Mesin Pertamina Enduro 4 Strok.* Jurusan Pendidikan Fisika. Universitas Negeri Yogyakarta

[5] Product Bulletin: Sodium Lauryl Sulfate. Stephan Company, Northfield, Illinois; 2012.

[6] Material Safty Data Sheet: STEPANOL WA-EXTRA K. Stephan Company, Northfield, Illinois; 2006.

[7] GuideChem. *Sodium Dodecyl Sulfate (CAS* 151–21–3*) MSDS.* Available at: http://www.guidechem.com/msds/151–21–3.html. Accessed August 19, 2015

[8] Dyck SV, Gerbaux P, Flammang P. Qualitative and quantitative saponin contents in five sea cucumbers from the Indian Ocean. Mar Drugs. 2010 Jan;8(1):173-89.

[9] Rowe, R. C., Sheskey, P. J., and Owen, S. C. 2006, *Handbook of Pharmaceutical Excipient. Edisi 5,* Pharmaceutical Press and American Pharmacist Association

[10] Tadros. Tharwat F. 2005. *Applied Surfactant: Principles and Application*, Wiley – VCH Verlag GmbH & Co, Weinhem.

IOP Conf. Series: Materials Science and Engineering 1071 (2021) 012010 doi:10.1088/1757-899X/1071/1/012010

Formulation and Physical Evaluation of Body Lotion Preparation of Kacip Fatimah (*labisia pumila*) Ethanolic Extracts as Antioxcidant

A H Kusumawati[1]*, A Munawaroh[1] and L Fikayuniar[1]
[1]Faculty of Pharmacy, Universitas Buana Perjuangan Karawang

E-mail : anggunhari@ubpkarawang.ac.id

Abstract. Kacip Fatimah (*Labisia pumila*) believed to have the antioxidant that played an important role in protecting against some diseases and delay the aging process. But, no research on kacip fatimah in his cosmetics formulation and efficacy to the skin. One form of cosmetic preparation is body lotion. Purpose: To determine the antioxidant activity of Kacip Fatimah ethanolic extract and to determine whether the concentration of Kacip Fatimah ethanolic extract affects the physical properties of the body lotion preparation. The body lotion is made in the form of an oil in water (m/a) and made in 4 formulas the concentration of the extract 0.5 %, 1%, and 2%. The physical test of body lotion extract of *Labisia pumila* is organoleptic, homogeneity, pH, viscosity, the spread, and the attaching, the results of testing in analysis using SPSS with Kruskal-Wallis and Mann-withney. The result showed that extract ethanol of *Labisia pumila* having strong antioxidant activity it is 87,752 ppm, and it can be concluded that variations concentration extract ethanol of *Labisia pumila* 0,5 %, 1%, 2 % affects the physical properties of body lotion. The greater concentration of extract, decrease the viscosity value with FI = 2863 cPs, FII = 1974 cPs, FIII = 1514 cPs, decrease the pH with FI = 6,72; FII = 6,37; FIII = 6,23, increase the spreadability with FI = 7 cm, FII = 7,93 cm, FIII = 8,03 cm, and decrease the attachability with FI = 1,33 sec, FII = 1,3 sec, and FIII = 1,01 sec.

1. Introduction

The skin is an organ system that covers the entire human body which has a function as body protection against exposure to foreign objects, so skin health needs to be maintained and protected. The skin is said to be damaged if it has signs in the form of wrinkles, cracks, dryness, dullness, and scaly. Skin damage can be caused by free radicals. A free radical is a molecule that is relatively unstable with the atom in its outer orbit having one or more unpaired electrons. Free radicals become stable when they bind to electrons from other molecules. Antioxidants are chemical compounds that can provide electrons for free radicals, so free radicals cannot cause skin damage because they are successfully suppressed. Humans do not produce more antioxidants for reserves in their bodies, so if there is excessive exposure to radicals, the body needs antioxidants obtained from outside the body [1].

Kacip Fatimah or *Labisia pumila* is a traditional herbal medicine that has been used by many generations of Malay women to facilitate childbirth and as a postpartum medicine. Kacip Fatimah is also commercially available as a health supplement that is claimed to prevent and treat diseases, especially those related to the stability of female hormones. Many studies have been conducted to identify bioactive or phytochemical compounds in Kacip Fatimah that have pharmacological activities [2]. The antioxidant content in Kacip Fatimah is believed to play an important role in protecting against several diseases and delaying the aging process. From a cosmetic point of view, research has shown Kacip Fatimah's ability to specifically protect the skin against photoaging based on its high antioxidant activity [3].

Besides, the water-soluble Kacip Fatimah (*Labisia pumila*) leaf extract has identified phenolic compounds in the form of gallic acid. Gallic acid (GA) is a natural antioxidant phenolic compound extracted from plants that are widely used in food, medicine, and cosmetics [4]. However, there has been no research on Kacip Fatimah in cosmetic formulations and its efficacy for the skin. One form of cosmetic dosage is Body Lotion. Body Lotion is a practical preparation, easy and fast to apply to the whole body. For skin that needs protection from free radicals by utilizing natural ingredients, namely, Kacip Fatimah which contains high antioxidants, a Physical Evaluation, and Formulation of Kacip Fatimah Extract Body Lotion (*Labisia pumila*) is made as an antioxidant, as a development of previous research results.

2. Material and Methods

2.1 Material

The materials used are Kacip Fatimah (Labisia pumila) ethanol extract, stearic acid, glyceryl monostearate, glycerin, triethanolamine (TEA), methyl paraben, propyl paraben, Oleum rosae, Aquadest, DPPH p.a, Vitamin C.

2.2 Methods

Body Lotion preparations are made in the type of oil in water (M / A), with the procedure that the ingredients of the oil phase (stearic acid, glyceryl monostearate, liquid paraffin, and propylparaben) are mixed accompanied by heating at 70° C with a hot plate until homogeneous to form preparation A. In different containers, the aqueous phase (TEA, methylparaben, glycerin, and aquadest) is mixed at the same temperature until homogeneous to form preparation B. After being homogeneous, the two preparations are mixed at 70° C to form preparation C. At temperature space or about $30\text{-}35^{o}$ C is added with Kacip Fatimah ethanol extract, stirring until homogeneous, then added with fragrance, namely oleum rosae, and stirring continuously for 1 minute to form Body Lotion.

Table 1. Formulation of Body Lotion Ethanol Extract of Kacip Fatimah (*Labisia pumila*)

Material	Consentration (% b/v)			Function	Level of Use (%)
	F1	F2	F3		
Ethanol Extract of Kacip Fatimah	0,5	1	2	API's	
Asam Stearat	4	4	4	Emulgator	1-20%
Gliseril Monostearat	3	3	3	*Co-emulsifier*	-
Trietanolamine (TEA)	0,5	0,5	0,5	*Alkalizing*	2-4%
Gliserin	5	5	5	Humectan	≤ 30%
Parafin Liquid	2	2	2	Stabilizer	-
Metil Paraben	0,1	0,1	0,1	Preservative	0,02-0,3%
Propil Paraben	0,05	0,05	0,05	Preservative	0,01-0,6%
Oleum rosae	q.s	q.s	q.s	Fragrance	-
Aquadest	Ad 100%	Ad 100%	Ad 100%	Solvent	-

a. Organoleptic Test

The organoleptic test was carried out by observing the body lotion of Kacip Fatimah ethanol extract, which included the color, odor, and texture of the body lotion when applied to the skin.

b. Homogeneity Test

The homogeneity of the lotion is observed using a magnifying glass or visually, and observing the presence or absence of coarse or inhomogeneous particles on the preparation.

c. Viscosity Test

The instrument used for the viscosity test was the Lammy viscometer with spindle No. 6 and a speed of 100 rpm.

d. pH test

The pH of the preparation is adjusted to the pH of the skin, which ranges from 4.5 - 7 and is determined using a pH meter [5].

e. Spreadability Test

The body lotion preparation of Kacip Fatimah extract was weighed as much as 0.5 g, placed in the middle of round glass with a scale, on top of the preparation another round glass that had been weighed was placed then left for 1 minute and the diameter of the spread was recorded. A load of 50 g is added to the cover glass and left to stand for 1 minute and then the diameter of the spread is recorded. The weight is added in multiples of 50 g to reach 200 g, then the diameter of the spread is measured.

f. Adhesion Test

Body Lotion with Kacip Fatimah extract weighed as much as 0.1 g, placed in the middle of the object-glass, and covered with another object-glass. The 50 g weights are placed on the lid for 5 minutes. The end of the cover slide and the lower end of the slide is attached to the clamp on the adhesion tester, then the load support is removed. The length of time the two slides were separated from the test instrument was recorded as the attachment time for the preparation.

g. Test the Antioxidant Activity of Kacip Fatimah Extract

The antioxidant activity test in this study was determined by the diphenyl picryl hydrazyl (DPPH) method by making a sample solution of kacip fatimah extract with methanol solvent with a concentration of 20 ppm, 40 ppm, 60 ppm, 80 ppm, and 100 ppm, the solution is taken as much as 2 mL and added with 2 mL of DPPH solution. The mixture is then homogenized and left for 30 minutes. The next step is measuring the absorbance at a wavelength of 517 nm and measuring the blank. The results of the antiradical determination were compared with vitamin C. The absorption value of the DPPH solution was calculated as percent inhibition (% inhibition) calculated by the formula [6]:

$$\% \mathrm{I} = \frac{\text{Abs control} - \text{Abs sample}}{\text{Abs control}} \times 100\%$$

Information :

I	= Inhibisi
Abs control	= Absorbance control
Abs sample	= Absorbance sample

3. Result and Discussion

3.1. Antioxidant Activity Test

The absorbance measurement of the sample was carried out at a concentration of 20 ppm; 40 ppm; 60 ppm; 80 ppm; and 100 ppm calculated after 30 minutes of incubation. The absorbance measurement was calculated at a wavelength of 517 nm. From the absorbance results, the inhibition-concentration relationship curve is obtained as shown in the following figure.

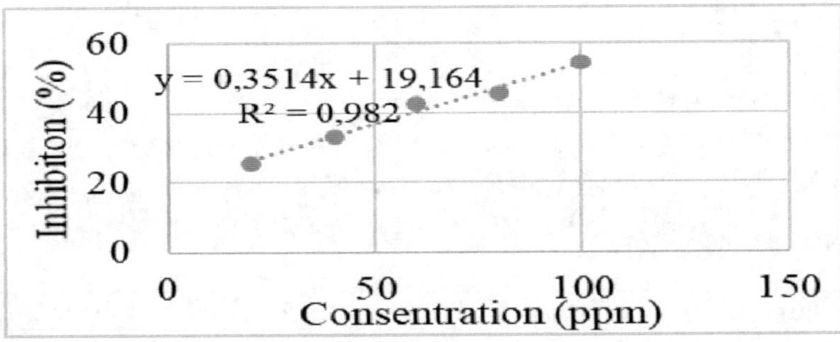

Figure 1. Relation of concentration (ppm) of sample with % inhibition

The IC_{50} value from the calculation when the% inhibition is 50% with the equation y = ax + b is 87.752 ppm. After knowing the IC_{50} value of the ethanol extract of Labisia pumila, the IC_{50} Vitamin C was measured with the following results.

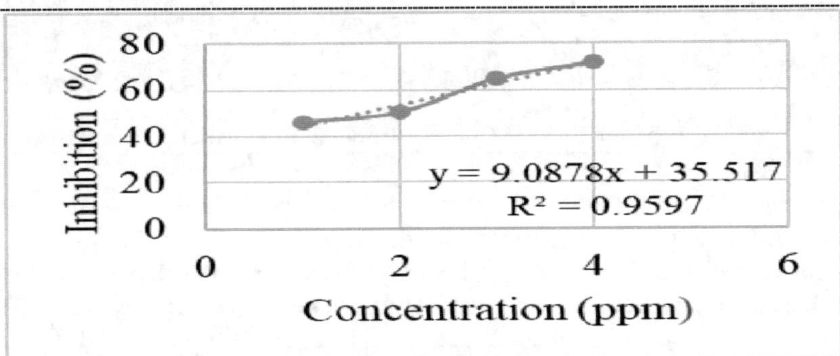

Figure 2. The relationship between vitamin C concentration and % inhibition

The IC_{50} value of vitamin C from the linear regression calculation resulted in 1.595 ppm. This shows that the antioxidant activity of positive control (Vitamin C) is still greater than Kacip Fatimah (*Labisia pumila*), but both are still categorized as very strong antioxidants because of the value of IC_{50} < 50 ppm [6].

a. Preparation of Ethanol Extract of Kacip Fatimah Body Lotion and Physical Evaluation of the Preparation

The results of the 3 formulations of the ethanol extract of Kacip Fatimah Body Lotion can be seen in the following figure:

Figure 3. Ethanol Extract Body Lotion Kacip Fatimah Formula I

Figure 4. Ethanol Extract Body Lotion Kacip Fatimah Formula II

Figure 5. Ethanol Extract Body Lotion Kacip Fatimah Formula III

The preparation of Body Lotion with active substances was carried out using the ethanol extract of the leaves of kacip fatimah (Labisia pumila). In the preparation of this Body Lotion, the first step is to make an ethanol extract from the leaves of kacip fatimah using the maceration method, then three formulas are made with various extract concentrations, namely 0.5%, 1%, and 2%. The difference of the formula lies in the variation of the extract concentration without any difference in additives because the aim of this study is to see the effect of extract concentration on the physical properties of the preparation. The 3rd result is the formulation of Body Lotion with the ethanol extract of Kacip Fatimah.

b. Physical Evaluation of Preparations

Physical evaluation of body lotion preparations includes testing of physical properties, namely organoleptic, homogeneity test, pH test, viscosity test, spreadability test, and adhesion test. The physical evaluation of body lotion preparations from kacip fatimah extract organoleptically aims to see to what extent a product is acceptable for quality using human senses. The results showed that the preparation formed a body lotion with a distinctive odor of kacip fatimah and a thick texture in formula 1 which was slightly thick in formula 2 and slightly thick in formula 3, this was due to differences in the concentration of the ethanol extract of kacip fatimah used which made the texture of the preparation thicker with each increase in concentration. The results show that the color of the body lotion preparation is light brown in formulas 1 and 2, while formula 3 is dark brown, this is due to the different concentrations used which make the color of the preparation darker with each increase in concentration. The brown color is obtained because the Kacip Fatimah extract is brown so that it affects the color of the preparation.

Table 2. Homogenity Test

Test	Criteria	Results		
		F1	F2	F3
Homogenity	Homogeneous	Homogeneous	Homogeneous	Homogeneous

An emulsion is said to be homogeneous if there is no visible separation between the components that make up the emulsion. The results of the homogeneity test are based on table 4.6 that the body lotion preparations for kacip fatimah extract show homogeneous results because there is no visible separation of the components of the body lotion, and there are no coarse grains [7].

Table 3. pH Test Results

Test	**Criteria**	**Results**		
		F1	F2	F3
pH	4,5-7	6,72 ± 0,078	6,37± 0,078	6,23 ± 0,15

Normal skin pH is between 4.5-7 [8]. Lotions that have a too alkaline pH can cause the skin to become dry, whereas if the pH is too acidic it will cause irritation [9]. The test results of the body lotion for Kacip Fatimah (*Labisia pumila*) extract showed that the pH of the preparation was in the range of skin pH with results from 6.23 to 6.72. Based on table 3, the higher the concentration of the extract, the more acidic the pH is, this is because the pH of Kacip Fatimah extract is acidic as indicated by changing the blue litmus paper to red after being reacted with the extract. To stabilize the pH of the preparation, Triethanolamine (TEA) was used because it is alkaline, which has a pH of 10.5 [10]. In addition, the pH of the preparation is influenced by the concentration of stearic acid. Stearic acid reacts with TEA to produce a salt, namely triethanolamine stearate.

Table 4. Viscosity Test Results

Test	**Criteria**	**Results**		
		F1	F2	F3
Viskosity	2000-50000 cPs	2863 ± 42,71	1974 ± 21,38	1514 ± 31,97

Viscosity is the resistance to fluid flow that experiences friction between liquid molecules with one another [11]. The greater the viscosity of the fluid, the more difficult it is for a fluid to flow and it also shows that it is increasingly difficult for an object to move in the fluid [12]. The viscosity requirement of a skin moisturizer is 2000-50000 cPs. Table 4 shows that the viscosity of the preparations for formulation I met SNI standards, while formulations 2 and 3 did not meet standards. This is because the concentration of the extract can affect the viscosity, the acidic nature of the extract will decrease the consistency of the preparation because the emulsifying value of glyceryl monostearate will decrease if it is in an acidic atmosphere. In this study, a combination of 4% stearic acid and 0.2% triethanolamine was used as an surfactant and glyceryl monostearate as a co-emulsifier. The combination of stearic acid and triethanolamine will form an anionic surfactant by forming an alkaline stearic salt [13].

Table 5. Spreadability Test Results

Test	**Criteria (cm)**	**Results**		
		F1	F2	F3
Spreadability	5-7 cm	7±0,58	7,93±0,12	8,03±0,15

The spreadability of the lotion indicates the ability of the lotion to spread to the location of use when applied to the skin. The requirements for the spreadability of the lotion are 5 - 7 cm [14]. The results show that the dispersion power that meets the requirements is only formula I, while formulas 1 and 3 do not meet the requirements.

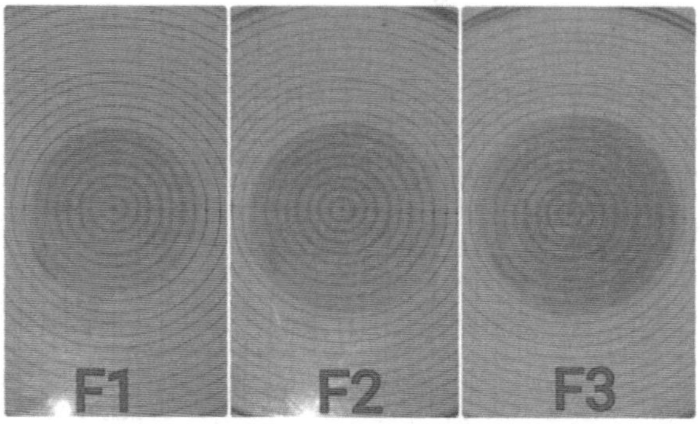

Figure 6. Spreadability Test Results

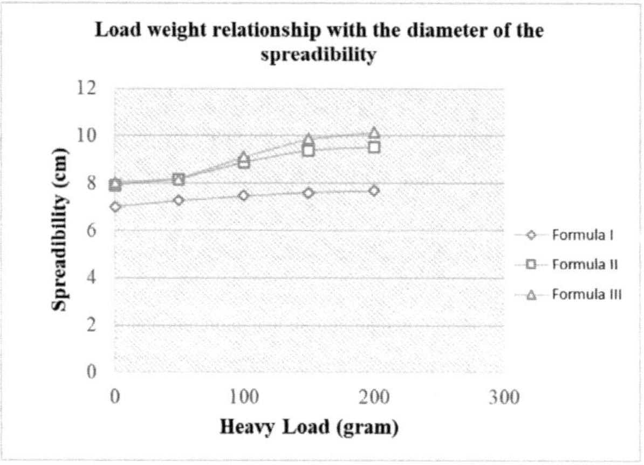

Figure 7. Graph of the Relationship of Increase in Load and Spread Power

The graph shows that the heavier the load added, the greater the diameter of the resulting dispersive power so that the increase in the diameter of the spreading power is directly proportional to the addition of the load. The preparation which has a high viscosity, namely formula I, has a diameter of dispersion in accordance with the test parameters. This is because the viscosity of the preparation affects the diameter of the spreadability of preparation is the viscosity. The lower the consistency of the lotion with the lower adhesion time, the easier it is to spread the lotion [15].

Table 6. Adhesion Test Results

Test	Criteria	Hasil		
		F1	F2	F3
Adhesion	> 1 second	1,33 ± 0,079	1,30 ± 0,02	1,01 ± 0,053

Based on the results shown in Table 6, the adhesion power of each preparation meets the adhesion requirements, namely> 1 second. The adhesion of preparation is influenced by the viscosity of the preparation itself. If the viscosity value is large, the adhesion will be high and if the viscosity is low, the adhesion will be smaller. Based on the results shown in Table 6, the adhesion power of each preparation meets the adhesion requirements, namely> 1 second. The adhesion of preparation is influenced by the viscosity of the preparation itself. If the viscosity value is large, the adhesion will be high and if the viscosity is low, the adhesion will be smaller. Based on the statistical analysis that has been carried out, the results of the Correlate test show a significance value of 0.395> 0.05, there is no correlation between viscosity and adhesion.

4. Conclusion

a. The ethanol extract of kacip fatimah (Labisia pumila) has a strong antioxidant activity, namely 87.752 ppm
b. Based on the research that has been done, the concentration of Kacip Fatimah (Labisia pumila) extract affects the physical properties of body lotion which includes organoleptic, pH, viscosity, dispersibility, and adhesion. Of the three formulas, Formula 1 with an extract concentration of 0.5% produced the preparation that had the best physical properties because it fulfilled all the body lotion requirements, namely with a pH of 6.72, a viscosity of 2863 cPs, a spreadability of 7 cm, and adhesion of 1.33 seconds.

References

[1] Mardikasari, S. A. et al. 2017. *Formulasi dan uji stabilitas lotion dari ekstrak etanol daun jambu biji (Psidium guajava L.) sebagai antioksidan*, Jurnal Farmasi, Sains, dan Kesehatan, **3**(2), pp. 28–32.
[2] Chua, L. S., Lee.,S.Y., Abdullah, N., Sarmidi, M.R. 2012. *Review on Labisia pumila (Kacip Fatimah): Bioactive phytochemicals and skin collagen synthesis promoting herb*, Fitoterapia. Elsevier B.V., **83**(8), pp. 1322–1335. doi: 10.1016/j.fitote.2012.04.002.
[3] Choi, H.K., Kim, D.H., Kim, J.W., Park, C.S., Sulaiman, N., Sarmidi, M.R. 2010. *Labisia pumila extract protects skin cells from photoaging caused by UVB irradiation*, Journal of Bioscience and Bioengineering. Elsevier B.V., **109**(3), pp. 291–296. doi: 10.1016/j.jbiosc.2009.08.478.
[4] Iwansyah, A. C. and Yusoff, M. 2013. *Identifikasi Dan Kuantifikasi Asam Galat Sebagai Sumber Antioksidan Pada Ekstrak Daun Kacip Fatimah (Labisia pumila) Larut Air*, Jurnal Aplikasi Teknologi Pangan, **2**(3), pp.133-138.
[5] Swastika, A, Mufrod, Purwanto., 2013, *Aktivitas Antioksidan Krim Ekstrak Sari Tomat (Solanum lycopersicum L.).* Trad Med Journal, **18**(3),132-140.
[6] Zuhra, C. F., Tarigan, J. B., Sihotang, H. 2008. *Aktivitas Antioksidan Senyawa Flavonoid Dari Daun Katuk (Sauropus androgunus (L) Merr.)*, Jurnal Biologi Sumatra, **3**(1), pp. 10–13.
[7] Erungan, A. C., Purwaningsih, S., Anita, S.B. 2009. *Aplikasi Karaginan dalam pembuatan skin lotion.* Jurnal Teknologi Hasil Perikanan Indonesia. 12 (2), pp.128-143.
[8] Barel, A.O., Paye, M., Maibach, H.I. 2001, *Handbook of Cosmetic Science and Technology.* Marcel Dekker, Inc. New York.
[9] Goskonda S. R., 2009, *Handbook of Pharmaceutical Excipients, Sixth Edition*, Rowe R. C., Sheskey, P. J., Queen, M. E. (Editor), London, Pharmaceutical Press and American Pharmacists Assosiation, 754-755.

[10] Febrianto T., Edi, S.S., Sunarno. 2013. *Rancang Bangun Alat Uji Kelayakan Pelumas Kendaraan Bermotor Berbasis Mikrokontroler*. Unnes Physics Journal, **2** (1), pp.30-34.
[11] Mutmainnah, Siti. 2008. *Pembuatan Counter waktu pada percobaan viskositas berbasis mikrokontroler HRS8000*. Skripsi. Universitas Islam Negeri Malang.
[12] Hendrawati, T. Y., Ambarwati, H., Nugrahani, R.A., Hasyim, U.H., Susanty, S. 2018. *The effects of adding aloe Chinensis baker gel extract on the formulation of lotion as a cosmetic product*, ARPN Journal of Engineering and Applied Sciences, **13**(23), pp. 9291–9297.
[13] Rowe, R. C., Sheskey, P. J., and Owen, S. C. 2006. *Handbook of Pharmaceutical Excipient. Ed. 5*. Pharmaceutical Press and American Pharmacist Association. New York.
[14] Dominica, D., and Handayani, D. 2019. *Formulasi dan Evaluasi Sediaan Lotion dari Ekstrak Daun Lengkeng (Dimocarpus Longan) sebagai Antioksidan*. Jurnal Farmasi dan Ilmu Kefarmasian Indonesia, **6**(1), pp. 1–7.
[15] Ansel, H. C., Popovich., Allen, L. V. 1989. *Pharmaceutical Dosage Forms and Drug Delivery System (Sixth Edition)*. Lippincott Williams & Wilkins. Philadelphia.

Physical Stability Test Sunscreen Gel Extracts Blackberry Fruit (*rubus fruticosus* l.)

N Gunarti[1*], I Aisyah[1] and F Lia[1]
[1]Faculty of Pharmacy, Universitas Buana Perjuangan Karawang

E-mail: neni.gunarti@ubpkarawang.ac.id

Abstract. Blackberry (*Rubus* sp.) fruit contains high levels of anthocyanins and other phenolic compounds, mainly flavonols and ellagitannins, which contribute to its high antioxidant capacity and other biological activities. Sunscreen preparations are cosmetic preparations that are used as protection to reduce the impact of sun exposure, whose formulations contain active ingredients to absorb or reduce sunlight, especially in areas exposed to ultraviolet rays and infrared waves. One of the potential natural ingredients for sunscreen is blackberry fruit (*Rubus* sp.). Contains flavonoid compounds that can prevent the harmful effects of UV rays. This study aims to determine the stability of the blackberry fruit extract sunscreen gel formula for 90 days of storage to determine the SPF value test results of blackberry fruit extract using UV-Visible. The research design is an experimental research conducted in a laboratory. The sample used in this study was Blackberry which was made into thick extract by maceration method. Furthermore, the thick extract obtained was made into a sunscreen gel and then tested for physical stability and the calculation of the SPF value. The sunscreen potency test is determined based on the method of calculating the percent erythema transmission value (%Te) and the percent transmission pigmentation (%Tp) as well as the Sun Protection Factor (SPF) value. On testing the SPF the average value on the test day 0, day 7, day 15, day 21, day 30, day 60, day 90, respectively 28,03±0,77; 28,28±0,16; 28,46±0,41; 28,66±0,59; 28,85±0,33; 29,04±0,11; 29.24±0,27. Percent of erythema (% Te) testing over 90 days was included in the total block category. Percent pigmentation (% Tp) testing for 90 days was included in the total block category and the SPF test for 90 days was included in the category ultra.

1. Introduction

Berries are well known for their properties that can improve the body's systems, antioxidants, vasodilators and antimicrobials [1]. Blackberry (*Rubus fruticosus* L.) is widely distributed in northern countries and has pharmacological activities, namely as an anticancer, antimicrobial, antioxidant, anti-dysentery, antidiabetic and anti-diarrheal [2].

Antioxidants are substances that can significantly inhibit or prevent oxidation caused by free radicals at small concentrations [3]. As for substances that can reduce the harmful effects of sunshine are sunscreens, sunscreens can ward off at least 85% of sunlight at a wavelength of 290-320 nm for UVB but can transmit light at a wavelength of more than 320 nm for UVA [4] Sunscreen is a cosmetic which physically or chemically can inhibit the penetration of UV rays into the skin [4]. Apart from chemical sunscreens that are also found in natural ingredients, sunscreens made from natural ingredients are widely used by people because they are considered safer. The ability to withstand UV rays from sunscreens was assessed by the Sun Protection Factor (SPF) [6]. Sunscreen in gel preparations that are made must meet stability requirements including physical stability tests, physical stability tests are carried out to ensure the quality, safety and benefits of the gel, meet the expected specifications and are stable during storage [7]. Research on sunscreen gel formulation has been carried out in research on the Formulation and Activity Test of Sunscreen Gel from Blackberry Fruit Extract (*Rubus fruticosus* L.) In Vitro with UV-Visible Spectrophotometry [8]. Therefore, the researchers are interested in conducting research on physical stability tests on the sunscreen gel formulation of Blackberry fruit extract (*Rubus fruticosus* L.) with accelerated stability for 90 days of storage.

2. Methods

2.1. Tools and Materials
The tools used in this research are Spatel, Watch Glass, Spreadability Test Glass, Stirring Rod, Beaker Glass, Oven, Erlenmeyer, Pipette, Water bath, Porcelain Cup, Measuring cup, Climatic chamber, UV-Visible Spectrophotometer (Thermo Scientific Evolution 201), Viscometer (Lamy Rheology First Touch 15.04T), pH meter (NeoMet pH-240L GJ-7726), Homogenizer (WiseTis HG-15D), Analytical balance (Adam PW254). The research materials used were Blackberry extract (*Rubus fruticosus* L.), Karbopol Blackberry extract (*Rubus fruticosus* L.), Karbopol 940, TEA, Nipagin, Nipasol, Glycerin, H_2SO_4, HCl, $FeCl_3$, Fragrance, Aquades, methanol (1.060009.2500).

2.2. Location and Time of Research
This research is an experimental research, which was conducted at the Laboratory of Preparation Technology, Faculty of Pharmacy, University of Buana Perjuangan Karawang in January 21 - May 21, 2020.

2.3. Experimental procedure
Experimental procedures include determination, preparation of raw materials, formulation of preparations, test of physical stability of preparations and test of SPF values.

2.4. Determination
Determination was carried out to ensure the correctness of the plants used in this study, this determination was carried out at SITH Labtek XI, ITB Bandung.

2.5. Phytochemical Screening
Phytochemical screening includes alkaloid test, flavonoid test, saponin test, tannin test and triterpenoid and steroid test.

2.6. Material Preparation
The part of the Blackberry (Rubus fruticosus L.) used was ripe fruit with a blackish purple color and made a thick extract using the maceration method.

2.7. Dosage Formulations
The formula of the preparation uses the active substance of Blackberry fruit ethanol extract, Karbopol 940 (gelling agent), glycerin (humectant), Triethanolamine (neutralizing agent), Nipagin (preservative), Nipasol (preservative), and aquadest (solvent).

2.8. Physical Stability
Physical stability tests carried out included organoleptic, preparation pH, viscosity, dispersion, and SPF value test.

2.9. Organoleptic

Organoleptic observations were carried out by observing changes in shape, color, and odor of the Blackberry fruit extract (*Rubus fruticosus* L.) sunscreen gel. The gel is usually clear with a semi-solid concentration [9].

2.10. pH

Observations were made on the gel preparation by dipping the pH meter into the gel preparation that had been made and then recorded. The pH of the skin is 4.5-6.5 [10].

2.11. Viscosity

The viscosity test is carried out to determine the thickness of the preparation that affects the time of use on the skin. A good viscosity of gel preparations is 2000 - 4000 Cps. [11].

2.12. Dispersibility

A total of 0.5 grams of the gel sample was placed on a round glass with a diameter of 15 cm, the other glass was placed on top of the gel and left for 1 minute. The scatter diameter is measured. After that, 150 grams of additional load were added and allowed to stand for 1 minute and then measured the diameter of the preparation. Good gel dispersion is 5-7 cm [11].

2.13. SPF Value Test

The SPF value is calculated by first calculating the area under the absorption curve (AUC) from the absorption value at a wavelength of 29-400 nm with 5nm intervals. The AUC value is calculated using the following formula (Chaims, 2004).

AUC = (Aa+Ab)/5x dPb-a

Information:

Aa = Absorbance at wavelength a nm

Ab = Absorbance at wavelength b nm

dPb-a = the difference between wavelengths a and b

The total AUC value is calculated by adding up the AUC value for each wavelength segment. The SPF value of each concentration is determined using the following formula: (Chaims, 2004).

Log SPF = AUC/(λn-λ1)

Information:

λn = The largest wavelength

λ1 = the smallest wavelength

3. Result And Discussion

The formulation of the Blackberry extract gel used a formula with a concentration of 1% Blackberry fruit extract.

Physical Stability Test

The physical stability test of the gel preparations was carried out by storing at 40 ° C with a relative humidity of 75% for 3 months using an oven. Stability testing was carried out for 90 days or 3 months, evaluated at 0, 7, 15, 21, 30, 60 and 90 days.

Organoleptic Test

Table 1. The result organoleptic test

Organolepic	Day test-						
	0	7	15	21	30	60	90
Shape	Semi Solid	semi solid	semi solid	semi solid	semi solid	semi solid	semi solid
Color	Red Brownish	Red brownish	Red brownish	Red brownish	Red brownish	Red brownish	Red brownish
Smell	typical weak berry	typical weak berry	typical weak berry	typical weak berry	typical weak berry	typical weak berry	typical weak berry

Table 1 states that the organoleptic test for the gel preparation did not show a significant change at each test and it can be said that the gel preparation is stable.

Ph

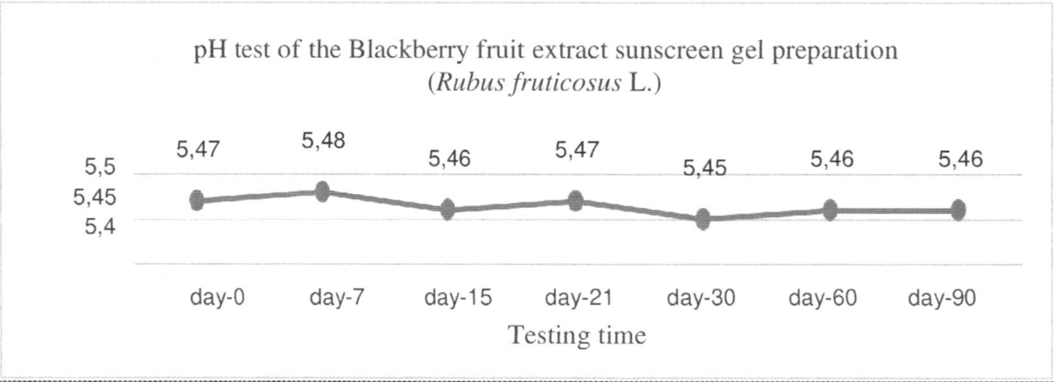

Figure 1. Graph pH test

In Figure 1, it states that the pH test for the gel preparation did not show a significant change in each test and it can be said that for the pH test using a pH meter for the sunscreen gel preparation of Blackberry extract (*Rubus fruticosus* L.) was stable for 90 days of storage.

Viscosity Test

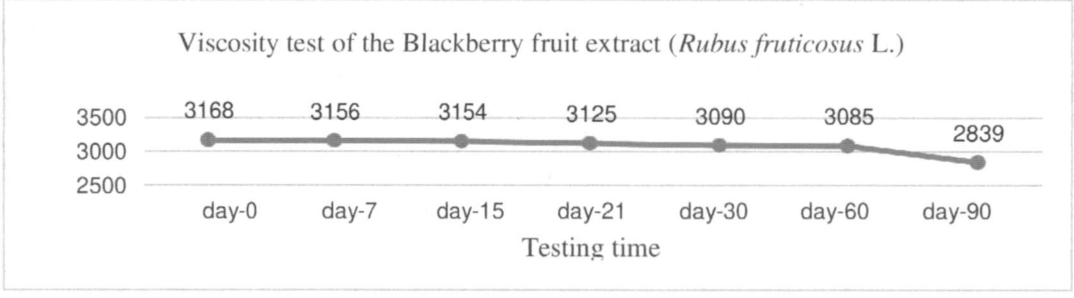

Figure 2. Graph Viscosity Test

In Figure 2 it states that based on the results of the viscosity measurement on the 0-90 day test, it has decreased. This can be due to the gel preparation showing its characteristics, namely synersis, which is the process of releasing the entangled liquid in the gel so that it allows the liquid to move towards the surface, therefore the preparation has decreased viscosity but is still in a good characteristic range (2000-4000cPs). The decrease in gel thickness can also be caused by external factors such as temperature and storage method.

Dispersibility Test

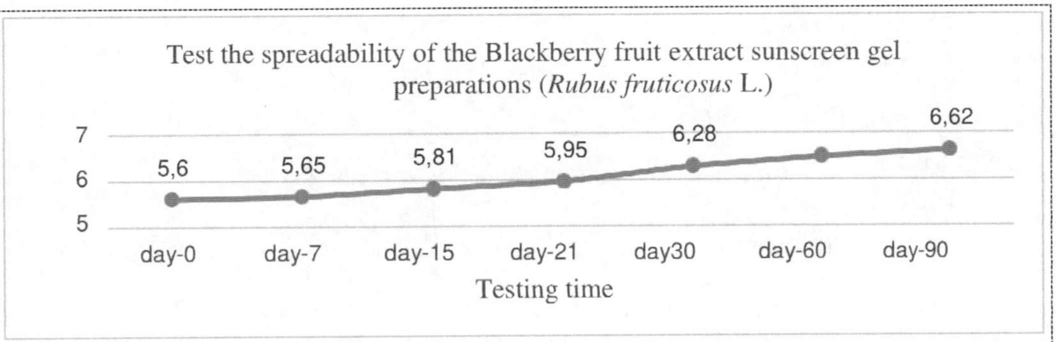

Figure 3. Graph Dispersibility Test

Figure 3 states that the results of the measurement of the dispersion power of the Blackberry extract (*Rubus fruticosus* L.) sunscreen gel, indicate that the gel has an increase in the diameter of the dispersion power.

SPF VALUE TEST

Percentage of Erythema Transmission

The Sun Protection Factor (SPF) value test was carried out in vitro using a UV-Visible Spectophotometer and the absorbance value was measured. The 10 g sample was dissolved in methanol up to 100 ml, then put in a 1 ml pipette into the cuvette and then the percent erythema transmission and pigmentation transmission percent (% Tp) were measured.

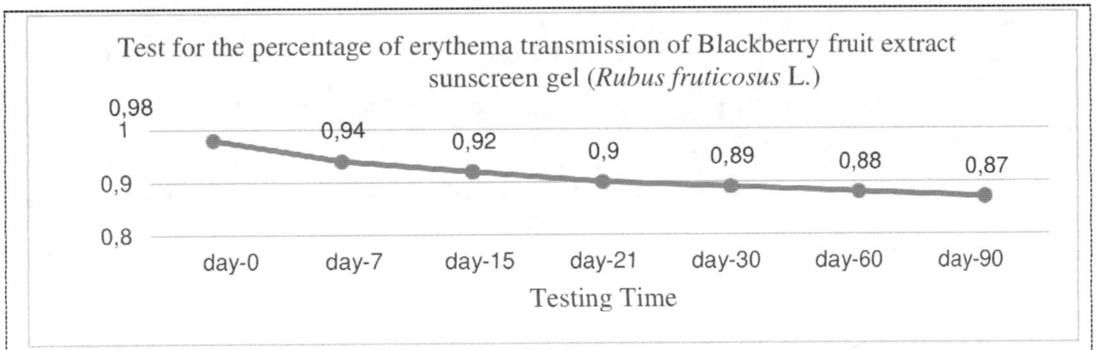

Figure 4. Graph of erythema percentage test

In Figure 4, it is stated that the test results for the percent erythema transmission (% Te) show that during the 90-day test, the percentage of erythema decreased, which means that the total flux transmitted by the sunscreen is getting less. This is because the gel experiences synergy. The value of% erythema as a whole did not change significantly and was still in the range % erythema <1% and included in the total block category.

Percentage of Pigmentation Transmission

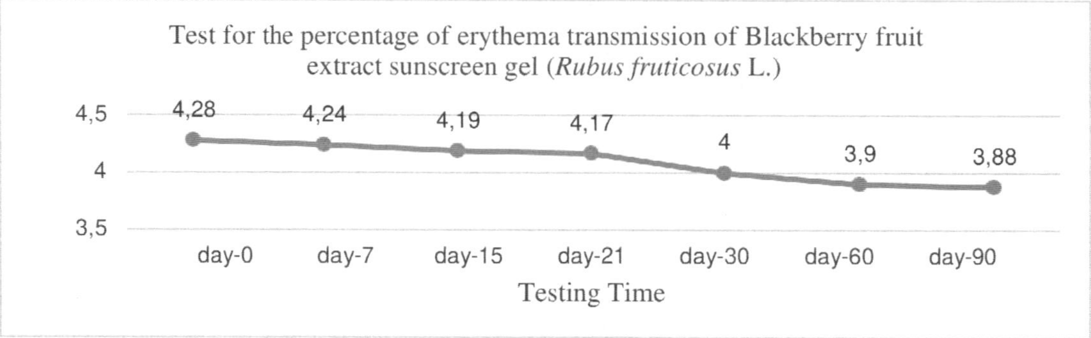

Figure 5. Graph of pigmentation percentage test

Figure 5 states that the results of the percent pigmentation test (% Tp) show that during the 90-day test, the pigmentation percent decreased, which means that the total flux continued by the sunscreen was getting smaller. This is because the gel experiences synergy. The overall pigmentation% value did not change significantly and was still in the range % 3-40% pigmentation and included in the total block category.

SPF Value Test

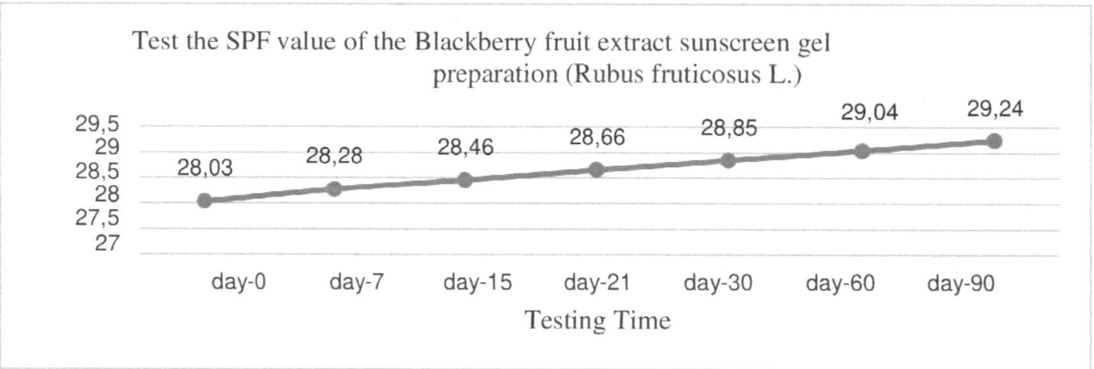

Figure 6. Graph of SPF Value Test

Figure 6 states that the results of measuring the SPF value carried out on the sunscreen gel preparation of Blackberry extract (Rubus fruticosus L.) stored for 90 days showed an increase in the SPF value. This is due to the fact that the gel can experience synergy, namely the occurrence of water discharge from the preparation so that it tends to remove water from the preparation (Kuncari et al, 2014). This resulted in decreased viscosity and increased levels measured (Martin et al, 1993). The SPF value resulting from testing for 90 days did not show a significant change so that the gel was still in the ultra and stable category during storage.

Data Analysis

Data analysis using Kruskal wallis analysis states that if the Asymp.Sig value <0.05, there is a difference from the tested viscosity so that the viscosity test decreases due to syneresis.

4. Conclusion

After doing research, physical stability test for 90 days of storage in an oven at a temperature of 40 °C and humidity <75% sunscreen gel extract of Blackberry (*Rubus fruticosus* L.) was stable during storage.

Suggestion

It is necessary to carry out chemical stability tests and long term stability tests at storage of 2 years 30 ± 2 °C with a humidity of 75 ± 5%.

References

[1] Bobinate R, Viskelis P, Venskutonis RR. 2012. *Variation of Total Phenolics, Antocyanins, Ellagic Acid and Radical Acavenging Capacity in Various raspberry (Rubus pp) cultivar.* Food Chemistry 132: 1495-1501.

[2] Dai J, Patel JD, Mumper RJ. 2007. *Characterization of Blackberry Extract and its Antiproliferative and Anti-inflamatory Properties.* Journal of Medicinal Food. 10: 258-256.

[3] Isnidar, Wahyuono, S., dan Seryowati, E.P . 2011. Isolasi Identifikasi Senyawa Antioksidan daun Kesemek (*Diopyros kaki* Thunb.) Dengan metode DPPH (2,2-Difenil-1 Pikrilhydrazyl). *Makalah Obat Tradisional.*

[4] Suryanto E. 2012. *Fitokimia Antioksidan.* Surabaya : Putramedika Nusantara.

[5] Shovyana, H,H dan Zulkarnain, A.K. 2013. Stabilitas Fisik dan Aktifitas krim tabir surya ekstrak etanol Buah Mahkota Dewa (*Phaleria macrocarpa* Scheff). Vol 18.

[6] Satiadarma, H dan Suyoto. 1998. Kesehatan Kulit dan Kosmetika. Yogyakarta : Andy Offset.

[7] Sayuti NA. 2015. Formulasi dan Uji Stabilitas Fisik Sediaan Gel Ekstrak Daun Ketepeng Cina (*Cassia alata* L). Jurnal Kefarmasian Indonesia. Volume 5

[8] Gunarti, N.S dan Lia F. 2019. Formulasi dan Uji Aktifitas Gel Tabir Surya Ekstrak Buah Blackberry (*Rubus fruticosus* L.) Secara In Vitro dengan Spektrofotometri UV-Visible. Karawang : Universitas Buana Perjuangan Karawang.

[9] Ansel, H.C. Pengantar Bentuk Sediaan Farmasi. 1998. Jakarta: Universitas Indonesia. Edisi 4. 105

[10] Nabaibaho, Olivia H.Paulina V.Y.Yamlean, Weny Winoyo. 2013. Pengaruh Basis Salep Terhadap Formulasi Sediaan Salep Ekstrak Daun Kemangi (*Ocimun sanctum* L) Pada Kulit Punggung Kelinci yang dibuat Infeksi *Staphyloccocus aureus*. Jurnal ilmiah Farmasi. Manado : UNSRAT. Vol 2

[11] Garg, A. Aggarwal, D, Gagr, S dan Sigla, A.K. 2002. *Spreading Of Semisolid Formulation*. An update Pharmaceutical Technology.

IOP Conf. Series: Materials Science and Engineering 1071 (2021) 012012 doi:10.1088/1757-899X/1071/1/012012

Formulation and Physical Stability Test of Facial Gel Wash Dragon Fruit (*hylocereus polyrhizus*) Peel Extract

N Yuniarsih[1*], **I Lenterani**[1] **and Farhamzah**[1]
[1] Universitas Buana Perjuangan, Karawang, Indonesia

Email: nia.yuniarsih@ubpkarawang.ac.id

Abstract. Cosmetics are a necessity in everyday life. Cosmetics are divided into two types, namely cosmetics for care (skincare) and decorative cosmetics (make-up). The facial wash is a form of cosmetic care to clean the face which can be in the form of foam, cream, and gel. Gel is a preparation that has good distribution, the good release of active substances, and is easily washed with water. This research is to formulate dragon fruit peel (*Hylocereus polyrhizus*) into a facial gel wash. Dragon fruit peel is a waste that is often ignored by the community. The dragon fruit peel was extracted and then made a facial gel wash formulation with three formulas, namely 8%, 9%, and 10% followed by physical stability testing. This study aims to formulate and determine the physical stability of the facial gel wash formulation of dragon fruit peel extract stored at three different temperatures, at room temperature, under the sun, and 40°C. The stability of the preparation was seen based on the results of organoleptic testing, pH, viscosity, and accelerated foaming before and after storage. The results of the physical stability test were statistically processed using the One Way ANOVA method. The results of this study showed that the three formulas, especially formula 1, were more stable at room temperature, under the sun, and 40°C.

1. Introduction

In the present era, cosmetics are an important need and cannot be removed and omitted. Cosmetic is a substance used to treat a parts of body from head to toe. There are two types of cosmetics, decorative cosmetics or makeup and cosmetics for skincare. There are various types of skincare, based on the shape and pattern, one of which is a facial wash.

Facial wash is one type of cosmetics used on the face to clean the face from sticking oil and dirt. Facial wash is usually made in various dosage forms, such as foam, cream, and gel. Gel or also known as jelly is a semisolid system consisting of a suspension of small inorganic particles or large organic molecules and is penetrated with a liquid [1].

The choice of gel dosage forms because it has several advantages, namely good distribution on the skin, good release of active substances, and easy washing with water [2]. The gelling agent is an important compound in gel making. One example is a carbomer. Carbomer is a good gelling agent used as a thickener, has a high viscosity, and produces a clear gel color [3].

In its development, cosmetics have been used from generation to generation by ancestors, using raw materials that exist in nature. However, currently a lot of cosmetics on the market are made from dangerous ingredients. Many consumers have returned to using cosmetics made from natural ingredients.

Dragon fruit peel is a waste that is rarely used. Usually, the skin is removed only the fruit is used. In this study, the dragon fruit peel was extracted to be used as an active ingredient in face wash preparations. The face wash preparations that have been made are then evaluated and tested for their physical stability.

The physical stability test is a test conducted to determine the stability of long-term storage of preparations and shelf life of preparations or expired date. According to the Head of the Indonesian Food and Drug Regulatory Administration, stability testing can be carried out in the long term and be accelerated. Long-term stability testing was carried out for 36 months at 30˚C, while accelerated stability testing was carried out for 3 months at 40˚C. According to [4], the stability test carried out is physical and chemical testing. The stability test was carried out at three different conditions, namely room temperature, under sunlight, and a temperature of 40˚C [5].

Based on the description above, the author has made a facial wash gel preparation from dragon fruit peel extract. Furthermore, the facial wash gel preparation from dragon fruit peel extract that has been finished is tested for physical stability in three different conditions, namely room temperature, under the sun, and a temperature of 40˚C. This aims to determine whether the preparations made can last for a long time after accelerated physical stability testing.

2. Material and method

The extracted dragon fruit peel is made into a facial wash gel preparation. The facial wash gel preparation from dragon fruit peel extract was tested for stability under three different conditions, namely at room temperature, under the sun, and a temperature of 40˚C. Stability testing was carried out for 90 days, observations that were seen from the stability test of the facial wash gel preparation of dragon fruit peel extract in the form of organoleptic observations (including color, smell, and shape), pH, foam power, and viscosity. Observations were made on days 0, 1, 7, 15, 30, 60, and 90. Furthermore, the results of the observations were statistically tested using the One Way ANOVA method followed by the Post Hoc Test.

Dragon Fruit Peel Extract Facial Wash Gel Formula

Table 1. The Formula for Facial Wash Gel Extract of Dragon Fruit Peel

Ingredients	Concentration			Function
	F1	F2	F3	
Dragon fruit peel extract	8%	9%	10%	Active ingredient
Brown rice	1.3%	1.3%	1.3%	Scrub
EDTA-4Na	0.1%	0.1%	0.1%	Chelating agent
Glycerin	2%	2%	2%	Wetting agent
SLS	2.5%	2.5%	2.5%	Foaming Agent
Propylene Glycol	1%	1%	1%	Preservative solvent
Methylparaben	0.2%	0.2%	0.2%	Preservative

Preparation of Facial Wash Gel for Dragon Fruit Peel Extract

The procedure for making the formula begins with the addition of EDTA-4Na and glycerin to the distilled water then stirring until it dissolves. Next step is adding the methyl paraben which has been dissolved in propylene glycol. The temperature was increased to 40C, Sodium Lauryl Sulfate (SLS) was added gradually until it was homogeneous. Then adding the dragon fruit peel extract, brown rice, citric acid, coloring, and fragrance, stirring until evenly distributed. The final step is the addition of the carbomer gradually, after blend well; TEA is added gradually while stirring until the preparation is completely homogeneous.

3. Results and discussion

Phytochemical Screening of Dragon Fruit Peel Extract

After the phytochemical screening was carried out, the dragon fruit extract was positive for alkaloid, flavonoid, phenolic, steroid, and saponin secondary metabolite compounds.

Table 2. Phytochemical Screening of Dragon Fruit Peel Extract

Testing	Reaction	Literature	Result (+/-)
Alkaloids	1 ml of sample + 2 ml HCl 2N (heated 5 minutes, filtered), filtrate + dragendorff	Orange solution	+
Flavonoids	1 ml of sample + 0.1 gram of Mg powder + 5 drops of 2N HCl	Yellow, orange or red color solution	+
Phenolic	2 ml of sample + 1 ml of methanol + + 2 drops of 1% solution	Yellowish green color solution	+
Steroids	2 ml of sample + 3 drops of concentrated HCl + 1 drop of concentrated	Yellowish green color solution	+
Saponin	2 ml of sample + 5 ml of distilled water + 10 drops of KOH (shake for 5 minutes, let stand for 10 minutes)	A constant foam formed within 10 minutes	+
Tannins	2 ml of sample (heated for 5 minutes) + 3 ml of 1% solution	Blackish blue solution	-
Annotation :	(+) result : the sample contains the test compound (-) result : the sampel does not contain the test compound		

Organoleptic Observations

Before storage for three months, the preparation is purple-brown has a characteristic aromatic odor, and takes the form of a gel. After being stored for three months, the preparations stored at room temperature did not change in a significant odor, shape or color. Preparations stored in the sun did not change in odor and shape, while the color after 15 days changes from purple-brown to brown and after 30 days changes to green. Furthermore, the preparations stored at 40°C did not change in smell and shape, while the color changed from purple-brown to brown after 15 days.

Color changing that occurs in facial wash gel preparations from dragon fruit peel extract that are stored in the sun and at temperature of 40°C occur because the secondary metabolites of anthocyanins, namely flavonoids in dragon fruit peel extract have been damaged due to exposure to direct sunlight which causes the occurrence of oxidation reactions repeatedly for 90 days. Storage at high temperatures will cause the extract to become unstable. Storage at high 40°C for a long time causes the purple color disappeared and turn into brown.

pH Testing

The pH of the dragon facial wash gel extract was stored for three months at room temperature for formulas F1, F2, and F3. Then tested statistically and the results showed no significant difference. Meanwhile, the pH of the facial wash gel preparation for dragon fruit peel extract was the most stable in the sun and at a temperature of 40°C, namely the F3 formula because there was no significant difference after statistical tests. However, the average pH test data shows a decrease in the pH value from day to day but is still as required, namely 4.5 - 6.5. The pH of the facial wash gel for dragon fruit peel extract is stable because it still meets the requirements

The decrease in pH value that occurs is due to storage time and temperature. The longer the storage time will cause physical changes of the preparation and the high temperature will accelerate the physical changes that occur [5].

Table 3. Results of pH Testing for Facial Wash Gel Extract of Dragon Fruit Peel

Dragon Fecial Wash Gel Formulation	Temperature	**0**	**1**	**7**	**15**	**30**	**60**	**90**
	Room temperature	5,99	5,96	5,95	5,95	5,92	5,89	5,87
F1	Under the sun	5,98	5,96	5,95	5,93	5,92	5,92	5,90
	Temperature 40°C	6,03	5,99	5,98	5,98	5,96	5,95	5,92
	Room temperature	5,99	5,95	5,92	5,87	5,86	5,84	5,83
F2	Under the sun	5,98	5,97	5,96	5,95	5,92	5,86	5,84
	Temperature 40°C	6,03	5,98	5,96	5,93	5,92	5,91	5,90
	Room temperature	5,99	5,97	5,96	5,95	5,93	5,89	5,88
F3	Under the sun	5,98	5,95	5,94	5,92	5,92	5,89	5,87
	Temperature 40°C	6,03	5,97	5,95	5,92	5,91	5,89	5,87

Foaming Power Testing

After storage for 90 days, the results of observations that were analyzed statistically showed less significant differences in F1 stored in 3 different conditions. The observations also show that F1 is more stable than F2 and F3.

The foaming rate is more related to the aesthetic value that is preferred by consumers who think that a good soap produces more foam (Purnama, 2006). The results of the high evaluation of the facial foam wash gel extract of dragon fruit peel for 5 consecutive minutes obtained a range of 1.00 - 5.00 with an average of 91.76%, the increase and decrease in the value of foam power was due to the shaking method when testing the foam power using manual method, not using tools that have standard speed and time that can be adjusted according to your needs, for example, such as a magnetic stirrer.

Table 4. Results of Foaming Power Testing of Facial Wash Gel Extract of Dragon Fruit Peel

Dragon Facial Wash Gel Formulation	**Temperature**	**Observation Day**						
		0 (%)	**1** (%)	**7** (%)	**15** (%)	**30** (%)	**60** (%)	**90** (%)
	Room temperature	91,99	92,83	93,41	93,54	93,61	92,47	92,33
F1	Under the sun	91,99	90,91	96,49	91,72	91,13	93,18	94,37
	Temperature 40°C	91,99	87,19	91,37	92,58	92,80	90,91	91,78
	Room temperature	93,64	91,75	91,75	90,71	96,37	92,89	94,93
F2	Under the sun	93,64	81,49	87,13	92,65	91,64	91,20	95,55
	Temperature 40°C	93,64	91,68	94,50	92,56	92,41	92,17	92,05
	Room temperature	85,59	90,29	91,10	91,30	92,95	96,10	93,58

F3	Under the sun	85,59	87,13	95,58	91,92	87,53	87,58	94,02
	Temperature 40°C	85,59	89,53	93,16	91,11	88,29	93,74	96,03

Viscosity Testing

There was an increase in the viscosity of the facial wash gel preparation of dragon fruit peel extract which was stored in 3 different conditions. Data processing that has been done shows that there are more significant differences in F2 and F3. In this test F1 is more stable when compared to F2 and F3.

The increase in viscosity of dragon facial wash gel preparations which are stored at room temperature, under the sun, and at a temperature of 40°C occurs due to the evaporation process caused by the storage container not being tightly closed so that the water content in the preparation is reduced which makes the preparation thicker. Storage in the sun and a temperature of 40°C further accelerates the evaporation process of water contained in the preparation which results in F1, F2, and F3 stored in the sun and at a temperature of 40°C which is thicker when compared to F1.

Table 5. Results of Viscosity Test for Facial Wash Gel Extract of Dragon Fruit Peel

Dragon Facial Wash Gel Formulation	Temperature	Observation Day						
		0 (Cp)	1 (Cp)	7 (Cp)	15 (Cp)	30 (Cp)	60 (Cp)	90 (Cp)
	Room temperature	2.279	2.386	2.478	2.615	2.712	2.726	2.721
F1	Under the sun	2.309	2.287	2.480	2.484	2.393	2.348	2.299
	Temperature 40°C	2.311	2.383	2.525	2.603	2.722	2.791	2.774
	Room temperature	2.110	2.081	2.278	2.382	2.421	2.407	2.422
F2	Under the sun	2.186	2.081	2.233	2.263	2.256	2.309	2.392
	Temperature 40°C	2.080	2.092	2.162	2.343	2.285	2.464	2.488
	Room temperature	2.229	2.492	2.686	2.721	2.794	2.820	2.804
F3	Under the sun	2.397	2.390	2.664	2.705	2.285	2.684	2.656
	Temperature 40°C	2.238	2.062	2.489	2.642	2.549	2.599	2.670

4. Conclusion

After testing, the conclusion of this study is that dragon fruit (*Hylocereus polyrizhus*) peel extract can be made into facial wash preparations with three different extract concentrations, namely 8%, 9%, and 10%. The results of the physical stability test of the facial wash gel preparation of dragon fruit peel extract were stable at room temperature, under the sun, and at temperature of 40°C. However, the best dragon facial wash gel preparation is F1 with an extract concentration of 8%.

Acknowledgments

We would like to thank Minister of Research and Higher Education, Republic of Indonesia, universitas Buana Perjuangan, for funding this study. Fauzi Akbar for supporting research experiments.

References

[1] Departemen Kesehatan Republik Indonesia. 2013. Farmakope Indonesia Edisi V. Jakarta : Kementerian Kesehatan Republik Indonesia

[2] Khoerina, Lulu. 2018. *Formulasi Gel Ekstrak Kulit Buah Naga Putih (Hylocereus undatus)*

Sebagai Antioksidan Dengan Variasi Gelling Agent Terhadap Sifat Fisik. Yogyakarta : Universitas Islam Indonesia.

[3] Rowe et al. (2009). Poloxamer: Handbook of Pharmaceutical Excipients, Sixth Edition. Handbook of Pharmaceutical Excipients, Sixth edition, 506–509.

[4] Kepala Badan Pengawas Obat dan Makanan, B. (2010). Petunjuk Operasional Pedoman Cara Pembuatan Kosmetik Yang Baik. 48(9), 1–16.

[5] Maio, Cosmetic Products Stability Guide, (2004). *Theme Series* (Vol. 1) ISBN-85- 88233-0.

IOP Conf. Series: Materials Science and Engineering 1071 (2021) 012013 doi:10.1088/1757-899X/1071/1/012013

Formulation and Antibacterial Activity Test of Foot Spray with Beluntas Leaf Ethanol Extract (Pluchea Indica L.)

Farhamzah[1*], A Herli[1] and I L P Mursal[1]
[1]Universitas Buana Perjuangan Karawang, Karawang, Indonesia

Email: farhamzah@ubpkarawang.ac.id

Abstract. Foot odor is one of the things that is quite disruptive to appearance, the 2014 American Podiatric Medical Association studied a significant increase in the prevalence of foot odor in the community. One of the causes of foot odor is the presence of bacteria, especially the bacteria Bacillus Subtilis. This study used the ethanol extract of beluntas leaf (*Pluchea Indica L.*) as the active ingredient because the ethanol extract of beluntas leaf has been widely used and proven to have antibacterial activity besides the ethanol extract of beluntas leaf has a high enough compatibility to be used as a cosmetic product, including spray preparations. The spray form was chosen because of its easy and pleasant use. This study aims to make a foot odor control product in the form of a spray with 3 concentrations of beluntas leaf ethanol extract, 0.5%, 1.0% and 1.5% to study the antibacterial activity using the well diffusion method. The results showed that the ethanol extract of beluntas leaf can be formulated into foot spray preparation with the results of the organoleptic examination of the three formulas are clear liquid, and yellow slightly green color. The pH of the preparation is in the range 5.10 - 5.92. In formula 3 (F3) with a concentration of 1.5% beluntas leaf ethanol extract, it has a strong bacterial inhibition zone with an average inhibition zone diameter of 12.5 mm.

1. Introduction

One part of the body that is very important for humans is the feet, both for women and for men, therefore feet need to be kept clean but sometimes they are often ignored, resulting in the appearance of odor problems on the feet. foot odor was found to be derived from isovaleric acid, which is produced when Staphylococcus epidermidis, a resident species of the normal cutaneous microbial flora, degrades leucine present in sweat. In addition, *Bacillus Subtilis* was detected in the plantar skin of subjects with strong foot odor, and this species was shown to be closely associated with increased foot odor [1]. *Bacillus Subtilis* is one of the normal flora species which is mostly found on the skin. The Bacillus SP bacteria group is known to have an important role in causing foot odor. *Bacillus Subtilis* was 11.5% in causing foot odor, while Staphylococcus epidermidis was 86.5% [1].

Many products have been produced for anti-odor, such as anti-bacterial bodyscrub and antibacterial soap. One of the most common cosmetics preparations in the market is the form of spray, which is used on the face, hands and feet. The spray form is chosen based on the nature of the spray which can provide a concentrated content, but at the same time it has a fast drying profile so that it is easy to use and pleasant to the user [2][3]. *Pluchea Indica* L. [4] is a source of phytochemical compounds, and antioxidants that can protect and prevent damage to cells against free radicals. The main antioxidants in *Pluchea Indica L.* [5] are triterpenoids, lignin glycosides, tannin, terpenes, including some flavonoids polyphenols, quersetin, quinik acid and derivatives eudesmane. *Pluchea.* based on the measurement of inhibition, the ethanol extract of beluntas leaf (*Pluchea Indica* L.) has anti-bacterial against Staphylococcus aureus, *Bacillus Subtilis*, and Pseudomonas aeruginosa. [6]. *Pluchea Indica* widely grown in Indonesia. [7].

This study aimed to determine that the ethanol extract of beluntas leaf can be formulated into foot spray preparation and to test effectiveness foot spray as an antibacterial against *Bacillus Subtilis* and it's stability test.

2. Material and Methods

Beluntas leaf was macerated extract with 96 % ethanol solvent concentrated with a Vacuum rotary evaporator until thick extract is obtained [8]. Thick extract was analyzed qualitatively phytochemical and used for produce of foot spray preparation. Phytochemical analysis conducted to know alkaloids, flavonoids, saponin, tannins, Steroid and triterpenoid. Foot spray is formulated using a base foot spray which is added a variety of Pluchea indica leaf extract. Foot spray was made with four formulas using extracts with different concentrations, as shown in table 1. Basillus subtilis obtain from the Laboratory of Microbiology, Faculty of Pharmacy, University of Buana Perjuangan karawang.

Table 1. Formulation

Material	Formula (%)			
	F0 (Blank)	F1	F2	F3
Beluntas Leaf Extract	-	0,5	1	1,5
Glycerin	0,2	0,2	0,2	0,2
Alcohol 96%	40	40	40	40
Menthol	0,5	0,5	0,5	0,5
Perfume	0,5	0,5	0,5	0,5
Tween 20	4,3	4,3	4,3	4,3
Aqua demineralisata added till	100	100	100	100

Evaluation of foot spray preparations is by organoleptic testing including color, odor, pH test using a pH meter, antibacterial activity test and stability test observed for 12 weeks at room temperature conditions.

The antibacterial activity test of foot spray preparation containing beluntas leaf extract was carried out against *Bacillus Subtilis* by using the well diffusion method for formulas one (F1), two (F2), three (F3) and negative control or without extract (F0). The test was carried out three times for all formulas. The materials used in the antibacterial power test were MHA (Mueller Hinton Agar), Mac Farland standard solution to make test bacterial suspensions, sterile distilled water as a solvent for testing bacteria and *Bacillus Subtilis* bacteria.

3. Result and Discussion

Based on the results of the phytochemical screening test, it was found that the ethanol extract of beluntas leaf was positive for alkaloids, flavonoids and tannins. Result of phytochemicals beluntas leaf extract (Pluchea Indica L.) is in table 2.

IOP Conf. Series: Materials Science and Engineering 1071 (2021) 012013 doi:10.1088/1757-899X/1071/1/012013

Table 2. Phytochemicals results of *Pluchea Indica* leaf ethanol extract.

Secondary Metabolite	Result
Alkaloid	+
Flavonoid	+
Saponin	-
Tannin	+
Steroid	-
Triterpenoid	-

Activities of flavonoids, moreover, are related to their ability to form complexes with proteins from the cell wall, which will result in damage to the permeability of the bacterial cell wall. Flavonoids have an antibacterial effect because of its ability to interact with DNA of bacteria [9].

There are four formulas of foot spray of beluntas leaf extract consisting of 0.0 % (F0), 0.5 % (F1), 1.0 % (F2), 1.5 % (F3). The results of the formulas showing there is relationship between concentration of the extract with color, the higher the concentration of extract (active ingredient), the color is getting darker. As shown in figure 1.

Figure 1. Foot spray of beluntas leaf extract. Description: F0 = 0.0 % Extract, F1 = 0.5 % Extract, F2 = 1.0 % Extract, F3 = 1.5 % Extract

The observation shows that foot spray containing extracts of beluntas leaf 0.5 % (F1), 1.0 % (F2) and 1.5 % (F3) have the same characteristics as F0, but have a yellow slightly green color (Table 3). The colors of F2, F3, and F4 were derived from the ethanol extract of beluntas leaf.

Tabel 3. Organoleptic Test

Formula	Organoleptic		
	Appearance	Color	Smell
F1	Clear, Liquid	Colorless	Floral fruity smell
F2	Clear, Liquid	Yellow slightly green	Floral fruity smell
F3	Clear, Liquid	Yellow slightly green	Floral fruity smell
F4	Clear, Liquid	Yellow slightly green	Floral fruity smell

IOP Conf. Series: Materials Science and Engineering 1071 (2021) 012013 doi:10.1088/1757-899X/1071/1/012013

The results of the pH examination, the pH range of the formula F0, F1, F2 and F3 was obtained between 4.58 - 5.92. The results of checking the pH of the four preparations are shown in Table 4.

Tabel 4. pH Test

Formula	pH
F0	4.58
F1	5,10
F2	5,92
F3	5,92

The degree of acidity (pH) of the formula without extract (F0), formula one (F1), formula two (F2) and formula three (F3) is still included in the skin pH range of around 4.5 - 6.5 [10]. This indicates that the pH of 5.10-5.92 in the foot spray with beluntas leaf extract is still in the pH range of the foot skin so it can be used safely for feet.

The results of the antibacterial test using the well diffusion method, it is found that all formulas provide the inhibition zone diameter and the results of the examination can be seen in Figure 2 and the inhibition zone can be seen in Table 5.

Based on the results of the research that has been carried out, formula 0 without active substances has antibacterial activity against Bacillus Subtilis in the weak inhibition category with an average diameter of 4.5 mm, in the inhibition area in Formula 1 with the amount of extract 0.5% has antibacterial ability against Bacillus Subtilis moderate inhibition category with an average diameter of the inhibition area of 6.7 mm, in the inhibition area in Formula 2 with the amount of extract 1.0% has antibacterial ability against Bacillus Subtilis moderate inhibition category with an average diameter of the inhibition area of 9.3 mm, and for formula 3 with a concentration of extract 1.5% had strong antibactery activity with a diameter of the inhibition area of about 12.5 mm [11].

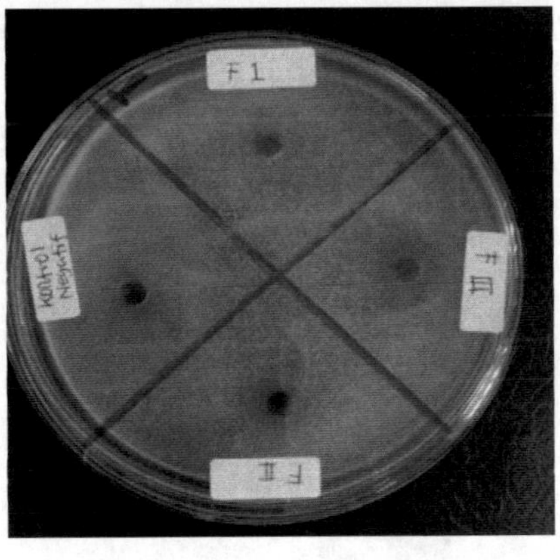

Figure 2. Antibacterial Activity Test

Tabel 5. Antibacterial Activity Test

Formula	Inhibition zone Dia.
F0	4,5 mm
F1	6,7 mm
F2	9,3 mm
F3	12,5 mm

The observations it can be seen that the increasing concentration of the extract also increases the inhibitory zone of the bacteria Bacillus Subtilis. Formula 0 without beluntas leaf extract provides 4.5 mm inhibition and shows a weak inhibition zone, and with the addition of the extract strengthens the inhibition zone of bacteria to a diameter of 12.5 mm for formula 3 with the addition of beluntas leaf extract was 1.5%.

Formula 0 without beluntas leaf extract provides a weak inhibition zone of 4.5 mm which can be caused by the ethanol content of 40% w / w in the formula. Alcohol is used in this formula as a solvent, provides a cool sensation to the preparation and makes the sprayed preparation dry on the feet quickly after spraying.

From the above results, formula three with a concentration of 1.5% beluntas leaf ethanol extract is effective in blocking the development of the *Bacillus Subtilis* bacteria which is a group of gram-positive bacteria. The strength of the extract in blocking bacterial development is due to the cell wall properties of the tested bacteria. Gram positive bacteria have a simpler cell wall structure that makes it easier for antibacterial compounds to enter the cells and reach their workplaces, while gram-negative bacteria have better resistance than gram-positive bacteria to antibacterial compounds [12]. Bacteria will die immediately if there is no cell wall, because they are unable to survive the influence of the surrounding environment [13].

Formulas 1, 2 and 3 that are produced are tested by putting them in a spray bottle, from the results of testing formulas 1, 2 and 3 can be sprayed, because formulas 1, 2 and 3 are clear liquids. Foot spray and spray bottle preparations can be seen in Figure 3.

Figure 3. Foot Spray Preparation

Product stability test results for 12 weeks at room temperature (28-30 oC), formula 1, 2 and 3 stable, clear liquid yellow slightly green color, pH 5.11-5.68 and floral fruity smell. Foot spray with beluntas leaf ethanol extract stabile and can be used for feet.

4. Conclusion

Beluntas leaf ethanol extract (Pluchea Indica L.) can be formulated into foot spray preparation and have antibacterial activity against Bacillus Subtilis in vitro. Formula 3 (F3) of foot spray on with the addition of 1,5 % Pluchea indica leaf extract is a formula most excellent and effective against Bacillus Subtilis and stable in storage for 12 weeks at room temperature (28-30 ^{o}C).

Acknowledgement

Thanks to Pharmaceutical laboratory of Faculty of Pharmacy, Buana Perjuangan University, Karawang, that have supported the research can be done.

References

[1] Ara K, Hama M, Akiba S, Koike K, Okisaka K, Hagura T, Kamiya T, Tomita F 2006 Foot Odor Due To Microbial Metabolism And Its Control *Can J Microbiol* **52(4)** 357-364

[2] Iswandana R and Sihombing LKM 2017 Formulation, Physical Stability, And In Vitro Activity Test Of Foot Odor Spray With Betel Leaf Etanol Extract (Piper betle L.) *Pharm Sci Res* **4**(3) 121-131

[3] Ashfia F, Adriane FY, Sari DP, Rusmini 2019 Formulasi Dan Uji Aktivitas Antibakteri Sediaan Footspray Anti Bau Kaki Yang Mengandung Ekstak Kulit Jeruk Nipis Dan Ampas Kopi *Indonesian Chemisry And Application Journal (Icaj)* **3**(1) 28-33

[4] Srimoon R and Suchanya N 2015 Antioxidant and antibacterial activities of Indian Marsh Fleabane (*Pluchea Indica* (L.) Less) *APST* **20**(2) 144–54

[5] Ahemd S A and Kamel E M 2013 Phenolic constituents and biological activity of the genus *Pluchea Der Pharma Chemica* **5**(5) 109–14

[6] Manu RRS 2013 Aktivitas Antibakteri Ekstrak Etanol Daun Beluntas (Pluchea indica L.) Terhadap Staphylococcus aureus, *Bacillus Subtilis*, dan Pseudomonas aeruginosa *Calyptra* **2**(1) 1-10

[7] Wiendarlina, I Y and Rizqiyana N 2019 Antibacterial Activity Roll On Deodorant With *Pluchea Indica* (L.) Leaf Extract Against *Staphylococcus Epidermidis* (Evans 1916) *In Vitro* The 2nd International Conference on Natural Resources and Life Sciences (NRLS) *IOP Conf. Series: Earth and Environmental* Science **293** 012031

[8] Ministry of Health Republic of Indonesia 2000 *Parameter Standar Umum Ekstrak Tumbuhan Obat* [General Standard Parameters of Medicinal Plant Extracts] (Jakarta: Direktorat Jenderal Pengawasan Obat dan Makanan) [in Bahasa Indonesia] p 1-12

[9] Pargaputri AF, Munadziroh E, Indrawati R 2016 Antibacterial effects of *Pluchea Indica* Less leaf extract on E. faecalis and Fusobacterium nucleatum (in vitro) *Dental Journal* **49**(2) 93-98

[10] Tranggono IR , Latifah 2007 Buku pegangan ilmu pengetahuan kosmetika. *PT. Gramedia Pustaka Utama.*

[11] Susanto D, Sudrajat and Ruga R 2012 Studi kandungan bahan aktif tumbuhan meranti merah (Shorea leprosula Miq) sebagai sumber senyawa antibakteri *Mulawarmnan Scientifie* **11**(2) 181-190.

[12] Maradona D 2013 Uji Aktivitas Antibakteri Ekstrak Etanol Daun durian (Durio Zibethinus L.), Daun Lengkeng (Zimocarpus Longan Lour) dan Daun Rambutan (Nophclium Lappaceum L.) Terhadap Bakteri Staphylococcus Aureus Atcc 25925 dan Escherichia Coli Atcc 25922, *Skripsi.*

[13] Fadhilla R 2010 Aktivitas Antimikroba Ekstrak Tumbuhan Lumut Hati (Marchantia palaecea) terhadap Bakteri Patogen dan Pembusuk Makanan. *Thesis.*

IOP Conf. Series: Materials Science and Engineering 1071 (2021) 012014 doi:10.1088/1757-899X/1071/1/012014

Chemical Characteristics of Motaf Sweet bread with Addition of Temulawak Starch

D Sandri[1] and E Lestari[1*]
[1]Agroindustrial Department, Tanah Laut State Polytechnic, South Kalimantan, Indonesia

E-mail: emalestari@politala.ac.id

Abstract. Bread is one of the processed foods from wheat which is widely consumed by the wider community which is used as a source of carbohydrates other than rice and noodles. In some countries, bread made from wheat flour is a staple food. Along with the development of the food industry made from wheat flour, the need for wheat flour is also increasing. So it is necessary to do efforts to find food sources other than wheat flour. In Indonesia, there are talipuk seeds which have the potential to be developed into a future source of carbohydrates and temulawak which contains quite a lot of starch. This study aims to analyze the chemical characteristics ofsweet bread *modified talipuk flour* (motaf)with the addition of temulawak starch. This research was conducted using a completely randomized design method (CRD) which consisted of 5 treatments consisting of the ratio of wheat flour, motaf, and temulawak starch flour. The sweet bread of these 5 treatments was tested for moisture content, ash content, protein, fat, carbohydrate and crude fiber analysis. The data were analyzed using the effectiveness test. The results showed that the best treatment produced on sweet bread made from the addition of modified talipuk flour (motaf) and temulawak starch was treated with the addition of 0.01 g of temulawak starch flour. The resulting sweet bread has 19.96% moisture content, 1.01% ash content, 8.65% protein content, 13.62% fat content, 56.76% carbohydrate content, 35.19% fiber content and the value of effectiveness results. the highest is 0.39.

1. Introduction

Bread is one of the processed foods made from wheat which is widely consumed by the public which is used as a source of carbohydrates other than rice and noodles. In some countries bread is a staple food. The advantages of bread include that it is easy to consume anytime and anywhere, nutritious and can be enriched with other nutrients so that it is good for children to adults and is also available in a variety of flavors. Sweet bread is a food made from wheat flour that is fermented with yeast bread and baked. The main ingredient in making bread is using wheat flour with a high gluten or wheat protein content. Flour is the basic ingredient used in making bread in various countries including Indonesia.

Indonesia is a country that has very diverse biological resources, one of which has the potential to be used as a food source. However, Indonesia is not a wheat producing country [1], while the demand for wheat flour in Indonesia is very large every year, this imbalance between demand and availability makes Indonesia have to continue to import wheat flour. Various things have been done to reduce imports by exploring local commodity materials that can be substituted as an alternative to wheat flour, including using cassava to be processed into flour called mocaf flour [2], utilization of gadung tubers into flour [3] and utilization. seaweed into flour [4]. One of the local commodity ingredients that can also be used as flour is from the cereal class, including lotus seeds which are called talipuk in the local language in one of the regions in Indonesia.

Talipuk is one of the plants used by the people of Hulu Sungai Utara, South Kalimantan, Indonesia as a source of food. One part that is used is talipuk seeds. Talipuk seeds can be converted into *Modified Talipuk Flour* (MOTAF) with high protein content and have met the Indonesian National Standard for wheat flour [5]. MOTAF can be used as a base for making various cakes, one of which is sweet bread. The research result states that MOTAF is able to substitute wheat flour by 5% in making sweet bread [1]. The sweet bread made in this research is not only from modified talipuk flour substitution (*motaf*) but also other ingredients added to increase the functional food value of the sweet bread, namely by adding temulawak starch flour.

Temulawak is a tuber plant that grows in tropical forest areas and has good health benefits. Temulawak also contains a large amount of starch so that it has the potential to be a commodity source of food to substitute for wheat flour, such as in the manufacture of sponge cakes that can substitute wheat flour up to 100% [6]. Based on these data, there has been no research that examines the substitution of motaf and temulawak starch in making this sweet bread and examines its chemical characteristics, it is necessary to conduct research on the chemical characteristics of sweet bread substituted for *modified talipuk flour* (motaf) and temulawak starch. This study aims to produce good chemical characteristics for motaf-based sweetbreads and curcuma starch so that it can be an alternative flour to substitute wheat flour from local Indonesian ingredients as a food source.

2. Method Material

Materials used in this study were curcuma, *modified Talipuk Flour (Motaf)*, wheat flour, chicken eggs, salt, sugar, yeast, *breave improver*, clean water, powdered milk, *margarine*.

Tools

Tools used in this study were a knife, grater, sieve, pan, scale, bowl, spoon, *mixer*, large basin, tray, measuring cup, brush, tongs.

3. Research Design

This research was conducted using a completely randomized design method (CRD) which consisted of 5 treatments with 3 replications. The treatment consisted of the ratio of wheat flour, motaf, and temulawak starch as shown in.

Table 1. sweet bread *Motaf* Temulawak

Formulasi	**A**	**B**	**C**	**D**	**E**
Motaf (g)	0	5	5	5	5
Wheat flour (g)	100	95	94,99	94,94	94,90
Temulawak starch flour (g)	0	0	0,01	0,05	0,1
Milk powder (g)	9	9	9	9	9
Yeast (g)	6,6	6,6	6,6	6,6	6,6
Margarine (g)	60	60	60	60	60
Sugar (g)	120	120	120	120	120
Bread improver (g)	1	1	1	1	1
Egg (butir)	1	1	1	1	1
Water (ml)	150	150	150	150	150
Fine salt (g)	5	5	5	5	5

Sweet bread from these 5 treatments was then tested for moisture content, ash content, protein, fat, carbohydrates and crude fiber analysis. Furthermore, the data were analyzed using de garmo analysis, namely the effectiveness test.

3.1. Making of Temulawak Starch Flour

Temulawak peeled from the husk, washed thoroughly, grated then mixed with enough water and stirred until the water and grated temulawak are mixed. Then filtered using a tight sieve. Temulawak starch mixed with water is deposited, the starch that has settled is discharged and rinsed with new water while stirring to remove the sap included in the temulawak starch. Rinsing is done repeatedly until the rinse water looks clear and you get clean temulawak starch. Then discarded the last rinse water and dried the temulawak starch flour under the hot sun until dry [7].

3.2. Making Modified Talipuk Flour (Motaf)

Fermentation of talipuk seeds with the addition of tape yeast as much as 6% of the total number of talipuk seeds used. Talipuk seeds are fermented for 16 hours and covered. After the talipuk seeds have been fermented for 16 hours, the talipuk seeds are washed thoroughly. Then the talipuk seeds are dried in the sun for about 2 days until they dry out. Then the talipuk seeds are blended until smooth, then the flour is sieved using a sieve so that the resulting flour is soft and smooth. In order to produce flour that has the best quality, after sieving, the flour is dried under the sun for about 1 day [8].

3.3. Making Sweet Bread

Making bread products using the method [9] that is the first stage done by preparing starter materials such as milk, sugar, eggs and yeast. 150 ml of milk mixed with 120 grams of sugar then warmed, stirred until dissolved, and stopped heating. Put one egg in the mixture, stir until well blended, and make sure the liquid is cold, then add 6.6 grams of yeast and stir until evenly then let stand until the liquid thickens and becomes foamy (± 10 minutes). After the starter ingredients are ready, the starter is put in a 300 gram mixture of flour, *motaf* and temulawak starch (according to the combination of the treatment formula), 5 grams of salt and 1 gram of *bread improver*, then stir until well blended. *margarine is* 60 grams ofadded to the dough, then kneaded until smooth, and after being smooth, the dough is covered using plastic ureping for about 1 hour until the dough expands. The dough is then formed and left to stand in a pan that has been smeared with butter for ± 30 minutes so that the dough expands, then oven for 20 minutes at a temperature of 150 $^{\circ}$C.

3.4. Chemical Characteristics Chemical

Testcharacteristics test in this study was carried out by using the proximate test, namely testing the water content [10], ash content [10], fat content [10], protein content [10], crude fiber [10], and analysis of carbohydrate content method by difference [11].

4. Results and Discussion

The results of the proximate analysis are as shown in the table below.

Table 2. Results of Proximate Analysis of Sweet Bread MOTAF Temulawak

No	Treatment	Moisture content (%)	Ash content (%)	Protein content (%)	Fat content (%)	Carbohydrate content (%)
1	K1	19.58 ± 0.25	1.01 ± 0.00	10.14 ± 1.92	14.15 ± 0.78	55.12 ± 0.89
2	K2	20.74 ± 0.37	1.02 ± 0.01	10.89 ± 0.72	13.57 ± 0.01	53.79 ± 1.09
3	MT1	19.96 ± 0.23	1.01 ± 0.00	8.65 ± 0.34	13.62 ± 0.08	56.76 ± 0.48
4	MT2	20.02 ± 0.65	1.02 ± 0.00	10.09 ± 0.33	13.34 ± 0.36	55.54 ± 1.34

5	MT3	22.85 ± 2.34	1.02 ± 0.00	9.88 ± 0.65	13.52 ± 0.23	54.72 ± 3.23
	SNI Sweet Bread	Max. 40	Max. 3	-	-	-

4.1. Moisture content

Water is an important component in food ingredients that can affect the appearance, texture, taste of food and its nutritional value [12]. The results of the analysis of water content (Table 5) showed that the more temulawak flour was added, the water content was higher (MT3: 22.85). [13] stated that water content determines the durability of a food. Food that is high in water content will spoil faster than food with low water content.

4.2. Ash content

Content The ash content describes the amount of minerals that are not burned into volatile substances [14]. The results of the ash content analysis (Table 5) show that there is no significant difference in the ash content of the sweet bread for each treatment even with the addition of temulawak starch flour. According to [15], ash is an inorganic mineral that has a high enough resistance to cooking temperatures so that its presence in food can change, but tends to remain.

4.3. Protein content

Is a source of amino acids that contain elements of C, H, O, and N which are not owned by fat or carbohydrates. Protein is a food substance that is very important for the human body, because protein functions as a fuel in the body and is also a building and regulatory material. The results of the analysis showed that the treatment of adding temulawak starch flour had decreased protein content. The low protein content in sweet bread with the addition of temulawak starch is due to the temulawak starch flour that does not contain gluten. With a low gluten content in the substituted sweet bread substituted with curcuma starch, it is hoped that it can reduce allergy symptoms in people with gluten allergy (gluten intolerance).

4.4. Fat content

Is a food substance that can produce energy for the human body. Functional food products must have low levels of fat, because low levels of fat can increase the effectiveness of dietary fiber in reducing cholesterol in the blood. The results of the analysis (Table 5), show that the treatment with the addition of temulawak starch flour decreased fat content, where with 100% wheat teung it was 14.15%.

4.5. Carbohydrate content

CarbohydrateCarbohydrates are one of the organic macromolecules composed mainly of three elements, namely carbon (C), hydrogen (H), and oxygen (O) with the general formula (CH2O) n. Carbohydrates have an important role in determining the characteristics of food ingredients, such as color, taste, texture, and so on. The results of the analysis of carbohydrate content (Table 5) indicate that the carbohydrate content of bread with the addition of temulawak starch has increased compared to K2. The increase in carbohydrate content in TPT sweet bread was due to the addition of temulawak starch. Starch is the main source of carbohydrates in food.

4.6. Fiber content

(fiber) can be divided into(crude fibercrude fiber) and(dietary fiberdietary fiber).(Crude fibercrude fiber) is plant / plant fiber that is insoluble in acids (H2SO4 1.25%) and alkaline (NaOH 1.25%). Crude fiber value is lower than dietary fiber because H_2SO_4 and NaOH have a greater ability to hydrolyze food components compared to digestive enzymes.

Table 3. Results of Crude Fiber Analysis of Sweet Breads MOTAF Temulawak

No.	Treatment	Crude Fiber (%)
1	K1	17.64 ± 1.61
2	K2	15.64 ± 1.20
3	MT1	15.10 ± 0.36
4	MT2	14.44 ± 0.92
5	MT3Dietary	15.32 ± 1.58

Fiber content ranges from 2-3 times crude fiber. The main components of crude fiber are composed of cellulose, gum, hemicellulose, pectin and lignin [16]. Crude fiber content in bread with MOTAF substitution and curcuma starch flour is lower than that of 100% wheat flour. This is probably due to the fiber content in MOTAF and low-fiber Temulawak Starch.

References

[1] E. Lestari, D. Sandri, F. Fatimah, and U. Umaira, “Volume Of Adonan Growth And Sweet Bread Sensory Made From Modified Talipuk Flour (Motaf),” *J. Teknol. Ind. Has. Question. Vol*, vol. 24, no. 2, p. 113, 2019.
[2] FN Amanu and WH Susanto, "Making mocaf flour in Madura (study of varieties and planting locations) on quality and yield [IN PRESS JULI 2014]," *J. Food and Agroindustry*, vol. 2, no. 3, pp. 161–169, 2014.
[3] VI Rosmeri, BN Monica, and CS Budiyati, “Utilization of gadung tuber flour (dioscorea hispida etc.) and mocaf flour (modified cassava flour) as a substitute material in making wet noodles, dry noodles, and instant noodles, " *J. Technol. Kim. And Ind.*, pp. 246–256, 2013.
[4] A. Agusman, SNK Apriani, and M. Murdinah, "The Use of Eucheuma cottonii Seaweed Flour in Making Rice Analogs from Modified Cassava Flour (MOCAF)," *J. Postharvest and Biotechnol. To the sea. And Perikan.*, vol. 9, no. 1, pp. 1–10, 2014.
[5] Fatimah, E. Lestari, and D. Sandri, "Characterization of Modified Talipuk Seed Flour (Nymphaea pubescens Willd) Using Tape Yeast," in *Proceedings of the PATPI National Seminar 2017*, Nov. 2017, pp. 256–262.
[6] E. Lestari, F. Fatimah, D. Sandri, and R. Yuniarti, "Potential Of Temulawak Steel As A Food Institution Of Wheat Flour In Bolu Cake," *J. Food Science Dan Has. Question.*, vol. 3, no. 1, pp. 60–67, 2019.
[7] E. Lestari, F. Fatimah, D. Sandri, and R. Yuniarti, “Potential of Temulawak Starch as a Food Substitute for Wheat Flour in Sponge Cakes,” *J. Food Science and Has. Question.*, vol. 3, no. 1, p. 60, 2019, doi: 10.26877 / jiphp.v3i1.3610.
[8] Fatimah, E. Lestari, and D. Sandri, "Characterization of Modified Talipuk Seed Flour (Nymphaea pubescens Willd) Using Tape Yeast," *Pros. Semin. Nas. PATPI 2017*, no. November 2017, pp. 256–262, 2017.
[9] Umaira. Lestari, Ema., D. Sandri, fatimah., "Volume Of Adonan Development And Sweet Bread Sensory Made From Modified Talipuk Flour (Motaf)," *J. Teknol. Ind. Has. Question.*, vol. 24, no. 2, 2019, doi: /10.23960/jtihp.v24i2.113-120.
[10] National Standardization Agency *for Food and Beverage Testing*, vol. 01-2891–19. 1992.
[11] IR Yenrina, M. Si, DA Permata, S. Si, M. Si, and I. Cover, *Analysis Methods of Food Materials and Bioactive Components Reviewer:* Andalas University Press, 2015.
[12] R. Syarief and H. Halid, "Food storage technology," *Arcan Jkt.*, 1993.
[13] FG Winarno, "Food Chemistry and Nutrition: The Latest Edition," *Jkt. Gramedia Pustaka Utama*, vol. 31, 2008.
[14] AD Soediaoetam, *Science of Nutrition*. Jakarta: Dian Rakyat, 1996.

[15] NW Desrosier, "Food preservation technology," *Translator M Muljohardjo UI-Press Jkt.*, 1988.
[16] D. Muchtadi, NS Palupi, and M. Astawan, "Laboratory Guidelines for Biochemical and Biological Chemical Methods in Evaluating the Nutritional Value of Processed Food," *Bogor ID PAU IPB*, 1992.

IOP Conf. Series: Materials Science and Engineering 1071 (2021) 012015 doi:10.1088/1757-899X/1071/1/012015

Probiotic Activity and Antibiotic Sensitivity of Lactic Acid Bacteria Isolated from Healthy Breastfed Newborn Baby Feces

E Lase[1], A L Davidson[1], I N E Lister[1] and E Fachrial[1*]
[1]Department of Biomedical, Faculty of Medicine, Universitas Prima Indonesia, Medan, Indonesia

E-mail: fachrial_edy@yahoo.co.id

Abstract. Breast milk is a very nutritional food that helps with the production of good microbiota in the gut of an infant. Generally, two lactic acid bacteria (LAB), namely *Lactobacillus* and *Bifidobacterium*, are related to infant gut microbiota. LAB is often used as a probiotic agent, but some conditions required for every bacterium to be utilized for this purpose. This study aimed to investigate the acid and bile salt tolerance as a criterion of probiotic agent and antibiotic sensitivity of LAB isolated from the feces of a healthy breastfed newborn baby. The sample was obtained from the feces of the aforementioned subject (baby). The result showed that the total colony of LAB from the feces was 10.2×10^8 CFU/mL. In an acidic condition at pH 3, there was one isolate that had 100% viability in the first hour period, but all of them had 100% at bile salt tolerance. However, sensitivity test of the isolated LAB indicated resistance to 6 types of antibiotics namely Erythromycin 5 µg, Gentamicin 10 µg, Oxacillin 5 µg, Ofloxacin 5 µg, Amoxycillin 25 µg, and Cefotaxime 30 µg. The study showed that LAB isolated from healthy newborn baby feces have potency as a probiotic.

1. Introduction

Metabolite of carbohydrate fermentation is majorly produced by lactic acid which are a group of Lactic Acid Bacteria (LAB) gram-positive bacteria. They are cocci or rod shaped, non-spore forming, but do not generate catalase [1]. During lactate fermentation, LAB produces many compounds such as diacetyl, organic acids, hydrogen peroxide, and bactericidal proteins [2].

There are several main genera of bacteria that are included in the LAB, including; *Carnobacterium, Aerococcus, Enterococcus, Lactococcus, Lactobacillus, Leuconostoc, Weissella, Oenococcus, Pediococcus, Tetragenococcus, Streptococcus*, and *Vagococcus* [3]. Furthermore, *Lactobacillus* has a beneficial effect on health when ingested by humans. Therefore, it is now widely isolated, and such strains are called probiotics [4], also, this type of bacteria usually grow (adhesion) and make colonies in the human intestine. Probiotics play a role in fighting against and preventing the colonization of pathogenic bacteria in the intestinal wall [5]. Some conditions required for bacteria to be a probiotic are safety, tolerant of acid and bile salt, capable of making colonies in the gastrointestinal tract, producing antimicrobial substances, and maintaining viability [6].

Many studies have shown that probiotics have a beneficial effect on health by preventing the growth of disease-causing bacteria, reducing the rate of colon cancer, increase the immune response, and decrease serum cholesterol [7]. Besides, they able to treat various diseases such as diarrhea, allergy, and hypertension [8].

In neonates, intestinal microbiota has the most important role in metabolism, nutrition, defense against pathogens and immune functions. *Bifidobacterium, Lactobacillus,* and *Enterococci* are anaerobic bacteria that are normally related to infant gut microbiota [7]. Breast milk is very nutritious and helps to produce good microbiota in an infant's gut [9].

Exclusive breastfeeding of life for the first six months is highly recommended for newborns because it provides nutrients that are needed for healthy growth and development [9]. Oligosaccharides, carbohydrates, fatty acids, nucleotides, cytokines, immunoglobulins, immune cells, lactoferrin, lysozyme, and other immunomodulating are bioactive molecules found in breast milk [10]. All of these molecules fortify the breast milk with ability to develop infant immune systems and defense against gastrointestinal and respiratory infections, allergies and also reduces the risk of developing other complications, such as inflammatory bowel disease (IBD), diabetes, or obesity [11]. *Lactobacillus* and *Bifidobacterium* are intestinal microbiota that are transferred from mother-to-infant through breastfeeding [11].

Antibiotic is the opposite of probiotic [6], but currently in many countries, antimicrobial resistance has caused high morbidity and mortality [12]. Actually, genetic mutation in bacteria is the primary cause of the resistance which is increased when antibiotics are underused, overused, or misused [13]. Patients' incompliance with recommended treatment and demand, irrational use of antibiotics, and drug advertisement are factors that also influence the resistance response [13]. Probiotics are chromosomally encoded to ensure absence of antibiotic resistance genes that can be transferred from them to another organism [6]. Furthermore, they are able to inhibit the growth of pathogenic bacteria and that effect is expected to reduce the use and resistance cases of antibiotics. Therefore, this study aimed to investigate the probiotic activity and antibiotic sensitivity of LAB isolated from the feces of healthy breastfed newborn baby.

2. Materials and Methods

2.1. Sample Collection

The sample was obtained from feces of a healthy breastfed newborn baby. It was collected at the healthy infant room of Royal Prima Hospital Medan. Afterwards, it was stored in a sterile sample tube and then carried to the Biomolecular Laboratory of Universitas Prima Indonesia Medan for immediate processing.

2.2. Isolation of LAB from Healthy Breastfed Newborn Baby Feces

LAB was isolated using specific growth media namely MRS broth and agar. 1 mL of feces sample was added to 9 mL sterile MRS broth in a test tube and incubated for 24 hours at 37°C in anaerobic condition. Afterwards, serial dilution process was performed (10^{-1} to 10^{-7}) by adding 0.1 mL of the LAB culture into 0.9 mL sterile MRS broth contained in a 1.5 mL microtube and then homogenized. This is the first, but the process was continued until the seventh one using the same method as initially described by taking 0.1 mL of the previous dilution as the portion to be used for the next addition. Following the seventh dilution, the product obtained was inoculated in a petri dish containing MRS agar, then incubated for 48 hours at 37°C for bacterial growth, and the total colony of LAB was counted afterwards [5].

2.3. Acid Tolerance

Please LAB isolate was enriched on MRS broth for 24 hours at 37°C. Then, 1 mL of LAB culture was added into a sterile test tube that contained 9 mL MRS broth as a control. In the same tube, few drops of 5M HCl were added to reach pH 3. Furthermore, the mixture was homogenized and incubated for 4 hours and at every hour it was diluted to 10^{-4}, followed by inoculation of the MRS agar. Then, incubated for 48 hours at 37°C, while the survival rate of LAB colonies was calculated in percentage [6].

2.4. Bile Salt Tolerance
1 mL of LAB culture was added into a sterile test tube that contained 9 mL MRS broth as a control. 0.03 gram bile salt was added in the same tube, after which it was homogenized and incubated for 4 hours and was diluted to 10^{-4}, at every hour. Then, inoculated on MRS agar and incubated for another 48 hours at 37°C, and the survival rate of LAB colonies was calculated in percentage [6].

2.5. Antibiotic Sensitivity
According to Kirby-Bauer, the antibiotic sensitivity test of LAB was conducted by disc diffusion method. LAB isolate was enriched on MRS broth for 24 hours at 37°C. Afterwards, it was swabbed on the entire MRS agar surface using a sterile inoculation needle, and then, the antibiotic disc was positioned on the media. There are 6 types of antibiotics used in this test, namely Erythromycin 5 µg (E_5), Gentamicin 10 µg (CN_{10}), Oxacillin 5 µg (OX_5), Ofloxacin 5 µg (OFX_5), Amoxycillin 25 µg (AML_{25}), and Cefotaxime 30 µg (CTX_{30}). The petri dish was incubated at 37°C for 48 hours, while after 48 hours, the diameter of the antibiotic disc and inhibition zone were measured using a calliper. Interpretation of the result of this test is sensitive when the diameter of inhibition zone $\geq$ 21 mm, intermediate when the diameter of inhibition zone 16 – 20 mm, and resistant when the diameter of inhibition zone $\leq$ 15 mm [6].

3. Results and Discussion

3.1. LAB Isolation from Healthy Breastfed Newborn Baby Feces
After incubating for 48 hours at 37°C under anaerobic condition, the total colony of LAB from healthy breastfed newborn baby feces was 10.2 x 10^8 CFU/mL. The colonies were milky or bright cream colored, and large in size, that is >1.5 mm. The growth of LAB isolate on the MRS agar surface is shown in figure 1.

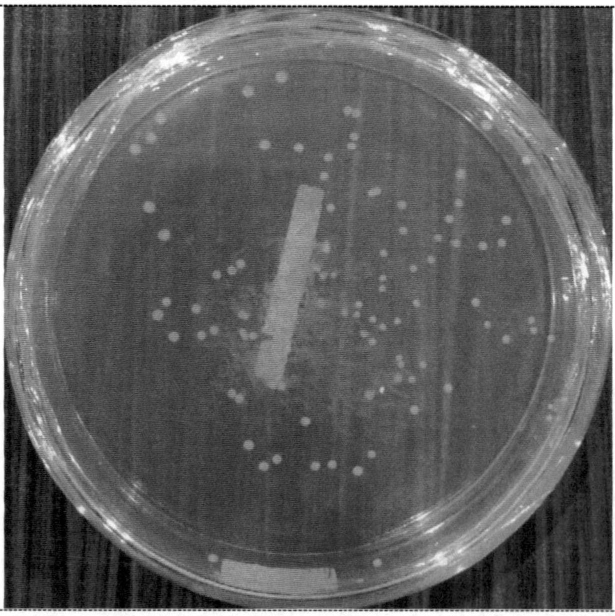

Figure 1. LAB growth from healthy breastfed newborn baby feces on MRS agar.

The minimum number of colony that a probiotic should have is 10^6 CFU/mL or gram, and to obtain it's beneficial effect, 10^8 to 10^9 of these microorganisms must be consumed daily [6]. Park Y S *et al* carried out the same study, which reported that the total colony of LAB isolated from newborn baby feces at Dongchimi was greater than 10^7 CFU/mL [14]. Also, another study that was conducted by Medjaoui I *et al*, stated that the average number of LAB colonies isolated from breast milk and infant's feces samples were 2.43 x 10^{10} CFU/mL and 6.85 x 10^{10} CFU/mL respectively [9].

3.2. Acid Tolerance of LAB Isolated from Healthy Breastfed Newborn Baby Feces

The growth of LAB isolated from healthy breastfed newborn baby feces on the MRS agar surface at pH 3 is shown in figure 2. Also, its tolerance test showed that only one isolate has 100% viability at pH 3 in the first hour period. The viability percentage of LAB from the feces in acidic conditions is shown in figure 3.

Figure 2. LAB growth on MRS agar in acidic conditions.

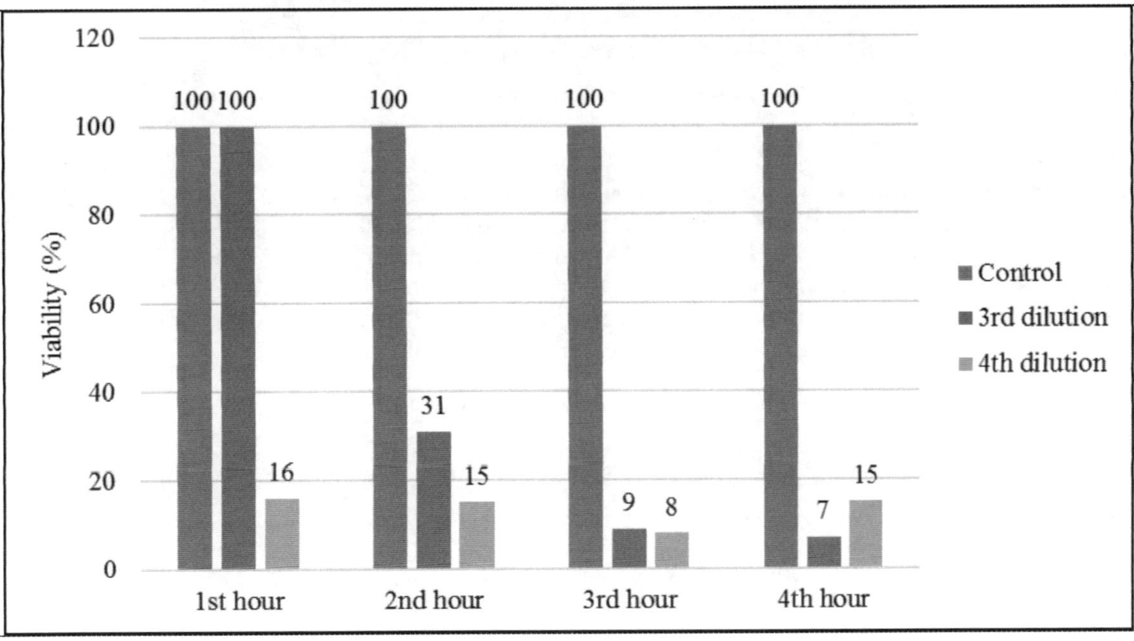

Figure 3. Viability percentage of Lab from healthy breastfed newborn baby feces at pH 3.

However, in order to assure viability, tolerance of acid conditions is one of the criteria that must be considered during the selection of potential probiotic isolates [15]. Same study by Park Y S *et al* also stated that LAB isolated from newborn baby feces at Dongchimi had the highest survival rate at pH 3 [14]. In addition, Adetoye A *et al's* study reported that LAB isolated from cattle feces were able to survive growth at pH 3 except four samples from *Lactobacillus* strains, namely *L.mucosae* C101, *L.ingluviei* C13, *L.ingluviei* C89, and *L.taiwanensis* C20 [16]. But, probiotic bacteria need to survive and stay in the stomach when in low pH conditions between 1.5 to 2.0 during the gastric emptying time of 4 hours or more [6].

3.3. Bile Salt Tolerance of LAB Isolated from Healthy Breastfed Newborn Baby Feces

Growth of the LAB on the MRS agar surface that contained 0.03 gram bile salt is shown in figure 4. All isolates from the feces have 100% viability of bile salt tolerance. Also, its viability percentage in the MRS agar is shown in figure 5.

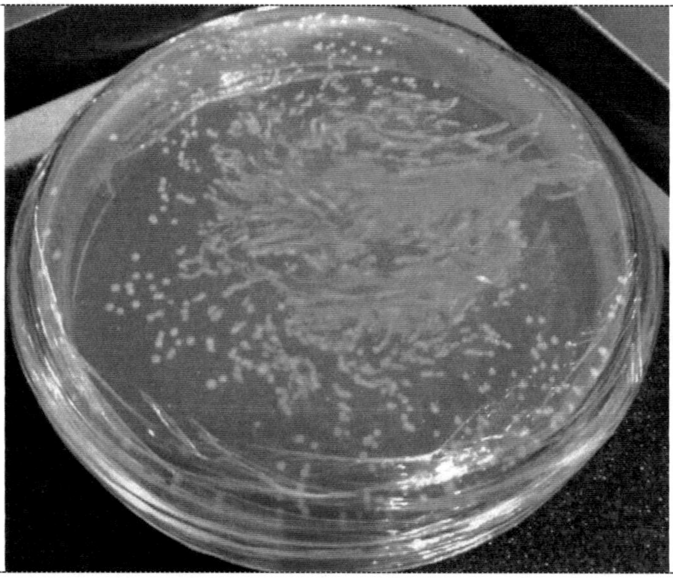

Figure 4. The growth of LAB on MRS agar that contained 0.03 gram bile salt.

Probiotics must have tolerance for bile salts to ensure colonization and bacteria metabolic activity in the host's small intestine. This is to assist Lactobacillus spp. and Lactococcus spp. balance the intestinal microflora and reach the small intestine and colon [17]. Furthermore, Park Y S et al reported that LAB isolated from newborn baby feces at Dongchimi were able to survive at 1% high concentration of bile [14]. Adetoye A et al's report also stated that LAB isolated from cattle feces survived at 5% bile salt concentration, while six other samples failed to grow at 7% [16]. Dissociation of a lipid bilayer and integral protein of cell membranes occurred after bacterial exposure to bile salts, which caused content leakage and finally, cell death [18]. Detoxification of the bile salt by producing bile salt hydrolase (BSH) enzyme activity was an ability of the probiotic strains [6]. Therefore, bile salt tolerance is related to the activity of the BSH, which hydrolyzes conjugated bile, thus minimizing the strains' bactericidal effect [19].

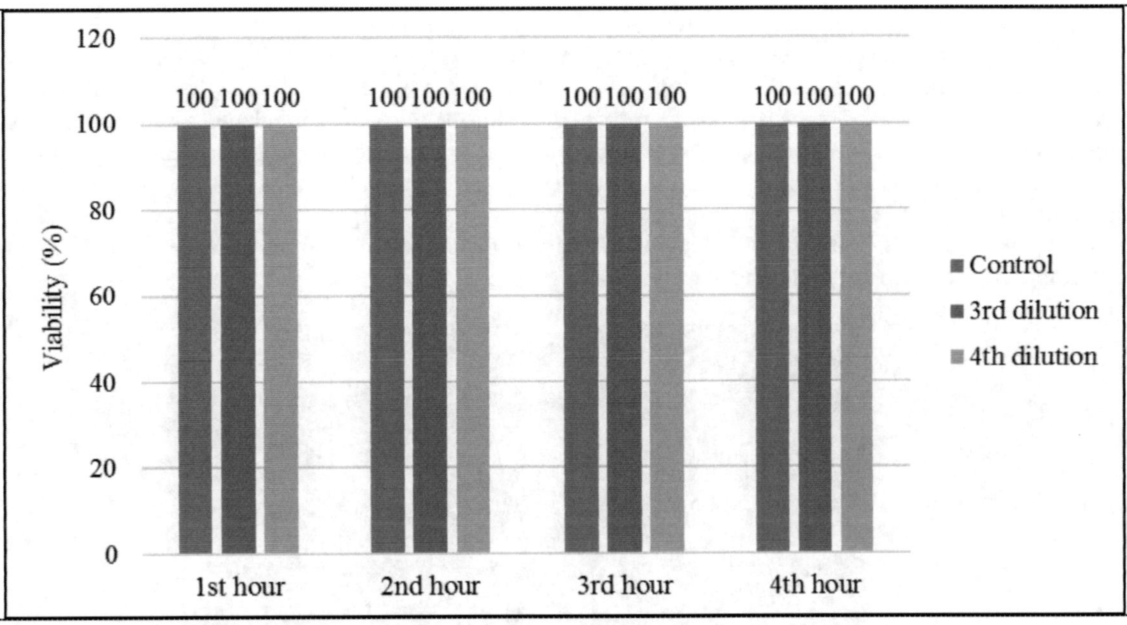

Figure 5. Viability percentage of LAB isolated from healthy breastfed newborn baby feces in MRS agar that contains 0.03 gram bile salt.

3.4. Antibiotic Sensitivity

The diameter of the antibiotic disc employed was 6.1 mm, and LAB was tested using 6 types of antibiotics (See Table 1). The LAB isolate was resistant with the diameter of inhibition zone ≤ 15 mm to all the antibiotics tested. The isolate's antibiotic sensitivity test can be seen in figure 6.

Table 1. Antibiotic sensitivity test of LAB from healthy breastfed newborn baby feces.

Isolate	The diameter of the inhibition zone (mm)					
	Erythromycin 5 µg (E_5)	Gentamicin 10 µg (CN_{10})	Oxacillin 5 µg (OX_5)	Ofloxacin 5 µg (OFX_5)	Amoxycillin 25 µg (AML_{25})	Cefotaxime 30 µg (CTX_{30})
F1	6.1	0	0	0	0	0

Therefore, transferrable antibiotic resistance determinants need not be possessed by potential probiotic strains [16]. From Adetoye A *et al's study,* 88 LAB samples isolated from cattle feces were resistant to Clindamycin, Vancomycin, Gentamycin, and Kanamycin but sensitive to Chloramphenicol, Ampicillin, Amoxicillin-Clavulanic acid, Erythromycin, and Tetracycline [16].

Generally, *Lactobacilli* strains were susceptible to the antibiotics such as clindamycin, chloramphenicol, erythromycin, and tetracycline, which all have mechanism of action that inhibits protein synthesis. But, the strains were more resistant to aminoglycosides such as streptomycin, neomycin, kanamycin, and gentamicin. *Bifidobacteria* strains were resistant to some nalidixic acid, namely metronidazole, polymixin B, vancomycin, gentamicin, kanamycin, streptomycin, colistin, trimethropine, and fusidic acid but susceptible to gram-positive spectrum (macrolides, novobiocin, teicoplanin, bacitracin, erythromycin, vancomycin, and lincomycin), broad-spectrum (rifampicin, chloramphenicol, and spectinomycin) and Beta-lactams antibiotics (piperacillin, amoxicillin, ampicillin, penicillin). Fundamentally, *Pediococci* strains had resistance to streptomycin, kanamycin, tetracycline, sulphamethoxazole, ciprofloxacin, doxycycline, trimethropine-sulphamethoxazole, and glycopeptides antibiotics such as vancomycin and teicoplanin but susceptible to penicillin G, imipenem, netilmicin, clindamycin, gentamicin, chloramphenicol, daptomycin, rifampicin, and ramoplanin. Also, *Leuconostoc* species were mostly susceptible to rifampicin, erythromycin, clindamycin, tetracycline, and chloramphenicol. Furthermore, these strains had resistance to glycopeptides antibiotics such as vancomycin, cefoxitin, and metronidazole and partially to gentamicin, kanamycin, streptomycin, nitrofurantoin, nalidixic acid, trimethropine, and sulphadiazine. *Lactococcus lactis* strains were susceptible to 1st generations of cephalosporin, erythromycin, chloramphenicol, amikacin, ampicillin, gentamicin, oxacillin, imipenem, vancomycin, penicillin, trimethoprim, and sulfamethoxazole. They also showed resistance to chloramphenicol, streptomycin, erythromycin, tetracycline, and clindamycin [20].

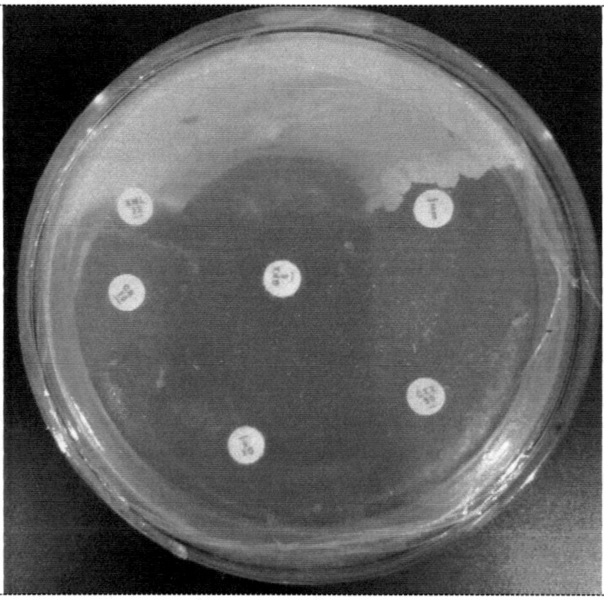

Figure 6. LAB isolate F1 antibiotic sensitivity test.

Enterococci indicated resistance to cephalosporins and low levels of aminoglycoside and clindamycin. Naturally, *Lactobacilli, Pediococci,* and *Leuconostoc spp.* all had high resistance to vancomycin and some of the *Lactobacilli* also had high resistance to bacitracin, cefoxitin, fusidic acid, ciprofloxacin, gentamicin, kanamycin, nitrofurantoin, metronidazole, norfloxacin, sulphadiazine, streptomycin, trimethoprim/sulphamethoxazole, teicoplanin, and vancomycin. In addition, *Lactobacilli* had a very high frequency of spontaneous mutation to nitrofurazone, kanamycin, and streptomycin [21].

Normally, probiotics are chromosomally encoded in order to prevent transferable antibiotic resistance genes [7]. Nevertheless, some reports showed that probiotics used in certain foods carry the transferable antibiotic resistance genes. Chromosomal, transposon, or plasmid located genes are possibly related to the occurrence of this resistance. The LAB has shown intermediate resistance, and susceptibility to various antibiotics. Therefore, international organizations such as European Food Safety Authority (EFSA) recommended microbiological breakpoint for the antibiotic to regulate the use of probiotic bacteria in foodusage. In fact, after antibiotic treatment, the microbiota in infant's gut are restorable by some probiotic strains with antibiotic resistance [22].

4. Conclusion

In conclusion, LAB isolated from healthy breastfed newborn baby feces have potential as a probiotic and the total number of their colony exceeded 10^6 CFU/mL. Also, they have good tolerance to acid and bile salt, which the viability reached 100%. And from the results of the antibiotic sensitivity test,

LAB were resistant (the diameter of inhibition zone ≤ 15 mm) to 6 types of antibiotics which include Erythromycin 5 µg (E_5), Gentamicin 10 µg (CN_{10}), Oxacillin 5 µg (OX_5), Ofloxacin 5 µg (OFX_5), Amoxycillin 25 µg (AML_{25}) and Cefotaxime 30 µg (CTX_{30}) tested.

Acknowledgments
The authors are grateful to Dr. dr. I Nyoman E Lister, M.Kes, AIFM, AIFO-K the head of Biomedical program, Edy Fachrial, M.Si head of Biomolecular Laboratory, Ditha Paramitha, and Michael Alfian Grey the analyst team of Medical Faculty in Universitas Prima Indonesia for their contribution to the success of this study.

References

[1] Fachrial E, Adrian and Harmileni 2018 *BioLink* **5**(1) 51-8
[2] Fachrial E and Harmileni 2017 *Technol. Sci. Am. Sci. Res. J. Eng.* **36**(1) 319–24
[3] Syukur S, Fachrial E and Jamsari 2014 *Res. J. Pharm. Biol. Chem. Sci.* **5**(6) 1096–104
[4] Ali F S, Saad O A O and Hussein S A 2013 *Egypt. Acad. J. Biol. Sci. G. Microbiol.* **5**(2) 21–34
[5] Syukur S, Yolanda S, Jamsari and Fachrial E 2015 *J. Chem. Pharm. Res.* **7**(9) 235–41
[6] Fachrial E and Harmileni 2018 *Res. J. Pharm. Biol. Chem. Sci.* **9**(6) 319–24
[7] Sun Y, Lou X, Zhu X, Jiang H and Gu Q 2014 *J. Bacteriol. Mycol.* **1**(2) 1–7
[8] Hendrati P M, Kusharyati D F, Ryandini D and Oedjijono 2017 *Biodiversitas* **18**(3) 1265–9
[9] Medjaoui I, Rahmani B, Talhi M, Mahammi F Z, Moghtit F Z, Mehtar N and Gaouar S B S 2016 *J. Pure. Appl. Microbiol.* **10**(4) 2613–20
[10] Turin C G and Ochoa T J 2014 *Curr. Trop. Med. Reports* **1**(2) 97–105
[11] Fernández L, Langa S, Martín V, Maldonado A, Jiménez E, Martín R and Rodríguez J M 2013 *Pharmacol. Res.* **69** 1–10
[12] Frieri M, Kumar K and Boutin A 2017 *J. Infect. Public Health* **10**(4) 369–78
[13] Odonkor S T and Addo K K 2011 *Int. J. Biol. Med. Res.* **2**(4) 1204–10
[14] Park Y S, Lee J Y, Kim Y S and Shin D H 2002 *J. Agric. Food. Chem.* **50**(9) 2531–6
[15] Menconi A, Kallapura G, Latorre J D, Morgan M J, Pumford N R, Hargis B M and Tellez G 2014 *Biosci. Microbiota Food Health* **33**(1) 25–30
[16] Adetoye A, Pinloche E, Adeniyi B A and Ayeni F A 2018 *BMC Microbiol.* **18**(1) 1–11
[17] Shehata M G, El Sohaimy S A, El Sahn M A and Youssef M M 2016 *Ann. Agric. Sci.* **61**(1) 65-75
[18] Hassanzadazar H, Ehsani A, Mardani K and Hesari J 2012 *Vet. Res. forum an Int. Q. J.* **3**(3) 181–5
[19] Ashraf R and Smith S C 2016 *Int. Food. Res. J.* **23**(2) 777–89
[20] Moračanin S, Djukić D, Zdolec N, Milijašević M and Maskovic P 2017 *J. Hyg. Eng. Des.* **18** 25–35
[21] Erginkaya Z, Turhan E U and Tatli D 2018 *Iran. J. Vet. Res.* **19**(1) 53–6
[22] Ramachandra B, Praveeen A R, Rajunaik B, Malashree L K and Prabha R 2017 *Int. J. Adv. Sci. Eng. Technol.* **5**(2) 33–8

IOP Conf. Series: Materials Science and Engineering 1071 (2021) 012016 doi:10.1088/1757-899X/1071/1/012016

Elastic Cross-Section Modulus Ratio of Jabon (Anthocephalus cadamba Miq.) Bolt-Laminated Timber Beams

Y A Pranata[1*], A Kristianto[1] and A Darmawan[1]
[1]Faculty of Engineering, Universitas Kristen Maranatha, Jl. Suria Sumantri 65, Bandung 40164, West Java, Indonesia

E-mail: yosafat.ap@eng.maranatha.edu

Abstract. The strength and stiffness of the beams are related to the elastic cross-section modulus of the beam, which are related to the cross-section moment of inertia. The elastic cross-section modulus of mechanically laminated (bolt-laminated) timber beams are not the same as solid timber beam for the beam with the same cross-section, this can be happened due to the inertia moment of cross-section of the bolt-laminated beam is lower than the solid beam due to slip between laminae and the elastic modulus of every lamina are varying. The aim of this research was to study the elastic cross-section modulus of mechanically laminated timber using bolts as a shear connectors. The scope of the study are the beam specimens made from Jabon timber (Anthocephalus cadamba Miq.) 60 mm x 160 mm cross-section size, three beam specimens with 1500 mm clear-span length, four lamina layers, lamination system using bolt as a shear connectors to reduce slip between laminae, flexural testing using the four-point loading test method according to the ASTM D198, the flexural behavior reviewed in this research are the flexural strength, the elastic cross-section modulus ratio, and the displacement ductility ratio. The bending strength of bolt-laminated timber beams obtained from an experimental test is 12.11 MPa (average) in the term of proportional limit load, while the bending strength of a solid timber beam is 37.96 MPa. The results showed that the elastic cross-section modulus and the flexural stress of the bolt-laminated timber beams were lower with a ratio of 0.32 than the solid timber beam, and the ductility ratio of the bolt-laminated timber beams was 1.18 so that they were categorized in the limited ductility criteria. The test results indicate that the failure of bolt-laminated Jabon timber beams are a failure of bending. The elastic modulus cross-section parameters of laminated timber beams are useful for the design of beam structure components in buildings, especially in the calculation of beam strength and beam stiffness as a serviceability requirement.

1. Introduction

Laminated timber, especially mechanically, has been widely applied to bridge structures and agricultural support building structures for a long time in America. The concept of mechanical lamination is that the wooden laminates are connected to a bolt joint, which acts as a shear barrier so that the slip that occurs between the laminates due to working loads can be prevented. The laminate system is one of the engineered wood solutions amidst the limited production of large diameter whole wood from forests. Mechanical laminated wood elements are structural components composed of horizontal or vertical laminates, which are mechanically joined using nails, screws, or bolts [1]. The bonding behavior between the laminae occurs as a result of the horizontal slip between the laminae being prevented and restrained by the sliding linkage device at certain intervals with the mechanical coupling.

The aim of this research was to study the elastic cross-section modulus of mechanically laminated timber using bolts as a connectors. The scope of the study are the beam specimens made from Jabon timber (Anthocephalus cadamba Miq.) 60 mm x160 mm cross-section size, three beam specimens with 1500 mm clear-span length, four lamina layers, lamination system using bolt diameter 10 mm as a shear-connectors, flexural testing using the four-point loading test method according to the ASTM D198 [2], the flexural behavior reviewed in this research are the flexural strength, the elastic cross-section modulus ratio, and the displacement ductility ratio.

This study is a continuation of previous research conducted by Pranata et.al. [3], namely the ratio of the elastic section modulus of mechanical laminated wood beams with mechanical joining devices. The research studied beams made from Acacia Mangium, Keruing, Meranti Merah, Mersawa, Nyatoh, and Durian timbers. Results obtained from previous research indicated that the elastic cross-section modulus ratio were ranged from 0.38-0.91. In general, the parameters of specific gravity, bolt diameter, and the ratio of the number of rows per bolt distance contribute to the elastic section modulus ratio of laminated beam-bolts.

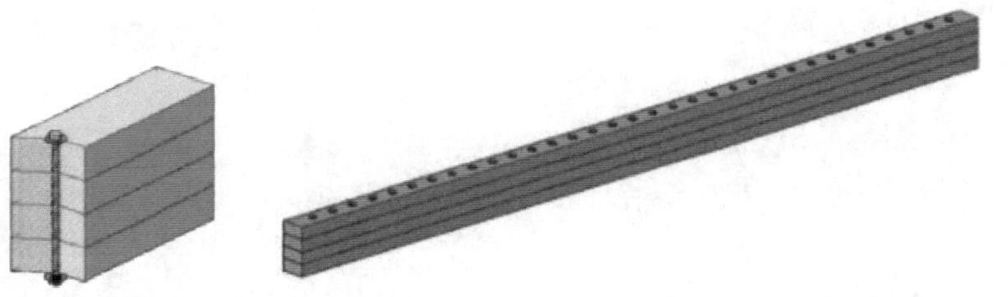

Figure 1. Schematic of bolt-laminated timber beams [4]

The research methodology is testing the test object experimentally in the laboratory. The study consisted of four stages, the first stage was a literature study on timber elastic modulus, timber flexural rigidity, and mechanical laminated timber. The second stage is the manufacture of mechanical laminated timber beam specimens using a bolt connection tool. The third stage is experimental testing in the laboratory. The fourth stage is a study to process the flexural capacity curve data, namely the load vs displacement relationship curve for the calculation of flexural rigidity, flexural strength, and displacement ductility ratio of mechanical laminated beams related to solid beams.

2. Basic Theory

2.1. Jabon Timber (Anthocephalus cadamba Miq.)

The Jabon tree (Anthocephalus cadamba Miq.) grows on various islands in Indonesia and is included in the strength-class III-IV category with a bending strength of 37.96 MPa, a modulus of rupture 67.79 MPa, and a modulus of elasticity of 6670.80 MPa. Jabon timber is included in the moderate class category of durability. Durability of timber as a building component against wood termite attacks, Jabon timber is included in the class II category, then resistance to wood rot fungus is included in the class IV-V category [5].

2.2. Elastic Section Modulus

The Elastic Section Modulus (S) in the elastic limit load range can be calculated by Equation 1 [6], which is based on the effect of the moment of inertia parameter of the cross section (I) and the distance from the outer edge fibre to the centre of gravity (y),

$$S = \frac{I}{y} \tag{1}$$

In the context of a mechanical laminated beam, the elastic section modulus is lower than that of a solid beam, the ratio can be calculated using Equation 2 and Equation 3 as follows,

$$S_{eff} = k_S \, .S \tag{2}$$

$$f_{B\text{-}lam} = \frac{M}{S_{eff}} \tag{3}$$

where M is the bending moment, Seff is the effective elastic modulus of the bolted laminated beam, kS is the ratio of the elastic section modulus of the laminated beam to the solid beam, and fB-lam is the flexural strength of the laminated beam [3].

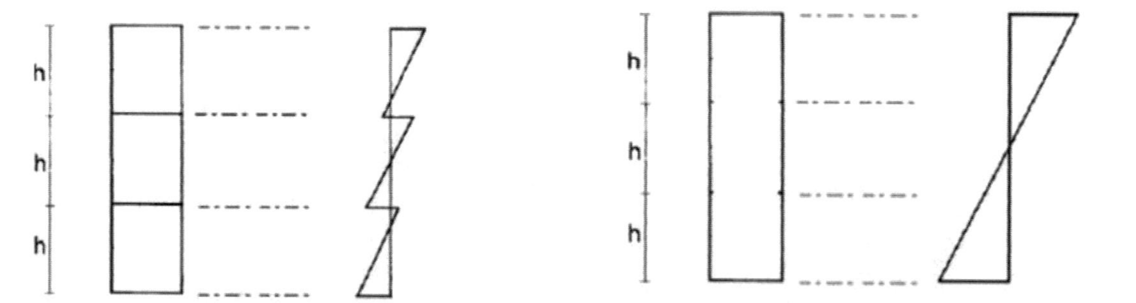

(a). Horizontally laminated beam (pranata et.al., 2012).

(b). Solid beam (Pranata et.al., 2020).

Figure 2. Cross-section of beam.

2.3. Deformation of Beam

The modulus of elasticity is the proportional slope of the linear line of the load and deformation curve (range of elastic load), in this context the bending load on the beam and deformation is the deflection of the beam due to the load acting. For a simple beam condition with a load of two centred on each distance as far as a from each support, the deformation relationship to the load is expressed in Equation 4 [6].

$$\delta = \frac{P.a}{24EI}\left(3L^2 - 4a^2\right) \tag{4}$$

With δ is deformation of beam due to flexural, P is concentrated load, L is length of beam span, E is modulus of elasticity (which is MoE parallel to the grain), and is the cross section moment of inertia. With the presence of more than one number of laminae, there is an interaction or slip between the laminates, because the compatibility of the strain mechanism between the bolt as a mechanical joint and lamina timber does not work perfectly and there is a partial interaction between the laminates, so that the bending behavior of the laminated beams is not the same as the solid beam (Figure 2.a and Figure 2.b). Equation 5 is the development of Equation 1 in the context of calculating the deformation of mechanical laminated beams, with the magnitude of the flexural rigidity of laminated beams being lower than those of solid beam, which is in the form of parameters for the flexural rigidity ratio of mechanical laminated beams (laminate-bolt) or k_{EI}.

$$\delta = \frac{P_{lam} a}{24(k_{EI}.E.I)}\left(3L^2 - 4a^2\right) \quad (5)$$

2.4. Method to Determine Proportional Point

The flexural capacity of timber beams is obtained from the results of experimental testing in the laboratory in the form of a bending load versus beam deformation curve. To obtain information on changes in the condition of the beam material from elastic to post-elastic, it is necessary to know the location of the proportional point. There are several methods to determine the proportional point of the previous research by Munoz et al. [7] One method of determining the proportional point is the Yasumura and Kawai method, with the basic principle that initial stiffness is calculated at conditions 0.1, 0.4, and 0.9 peak points (PU). The condition of the elastic material is determined by the intersection of two linear lines, namely the meeting point formed between the 0.1-0.4PU lines and the 0.4PU and 0.9PU lines. The meeting point is then shifted parallel to coincide with the load vs deformation curve of the experimental test results.

3. Experimental Tests and Discussion

In this study, a mechanical laminated timber beams were tested with a mechanical connectors, namely a 100 mm diameter bolts. Timber beams made of Jabon timber (Anthocephalus cadamba Miq.). The timber lamina has a cross-sectional size of 60 mm x 40 mm, the lamination system used is horizontal with 4 (four) laminae, so that the beam's final cross-section size is 60 mm x 160 mm. The distance between the bolts (spacing) is 100 mm.

Figure 3. Flexural test of the beam specimen using Universal Testing Machine.

Figure 3 show the testing process for one specimen of laminated timber beams. The test method is the four-point loading test with a distance of 500 mm (1/3 beam span). Furthermore, Figure 4 show the variety of beam failures that occur in the test object, namely simple tension, this is in accordance with one of the flexural test object failure models based on ASTM D143 [2] so that the expected failure occurs in accordance with the test results.

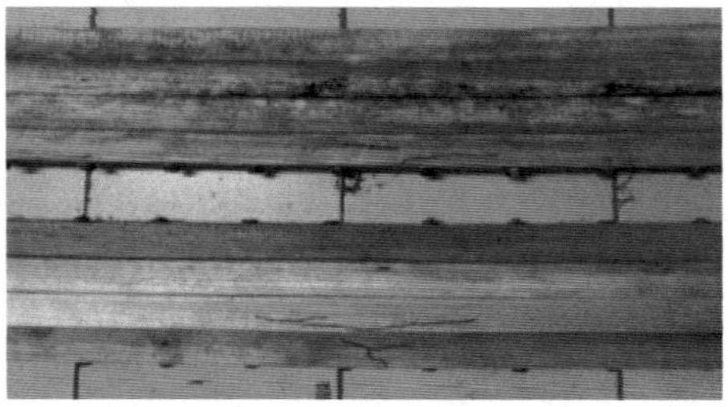

Figure 4. Failure type of the beam for all beams specimens which are simple tension.

Figure 5 show a curve diagram of the load versus beam deformation relationship. The test results of all specimens show almost the same curve trend, the maximum deformation ranges from 24.32 mm to 37.27 mm, while the maximum load ranges from 12.28 kN to 14.13 kN. Figure 6 show an example of calculating the proportional point (proportional load) and the ultimate point (ultimate load). The complete calculation results are shown in Table 1. The results of Table 1 show that the average beam turnover ductility ratio is 1.18. The average proportional limit load is 12.40 kN, the ultimate limit load is 13.58 kN. Furthermore, using Equation 3, the flexural rigidity ratio of laminated beams is calculated. The complete calculation results are shown in Table 2. Calculation of the elastic section modulus and bending stress is carried out using Equation 2 and Equation 3.

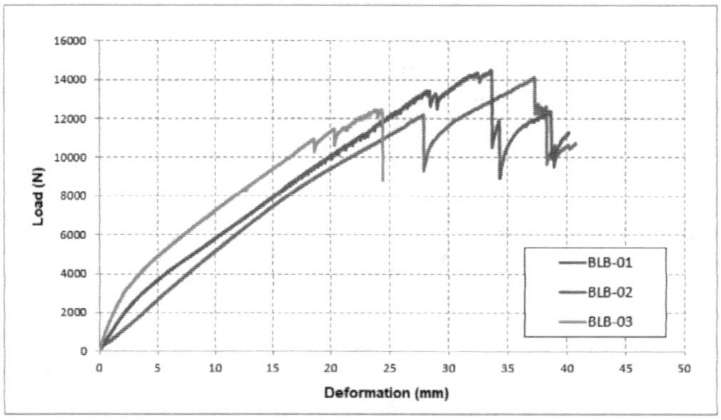

Figure 5. Load vs Deformation curves of the bolt-laminated timber beams.

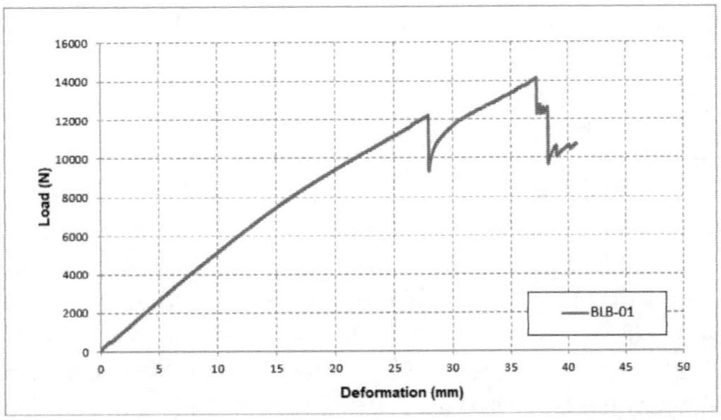

Figure 6. Calculation of proportional and ultimate points for specimen BLB-01.

Table 1. Proportional point, ultimate point, deformation at proportional point, deformation at ultimate point, and displacement-ductility results.

Specimen	P_y (N)	d_y (mm)	P_u (N)	d_u (mm)	μ
BLB-01	12001.91	23.83	12280.28	24.32	1.02
BLB-02	13074.70	28.46	14330.85	33.58	1.18
BLB-03	12137.70	27.78	14127.15	37.27	1.34
Average	12404.77	26.69	13579.43	31.72	1.18

Table 2. Results of flexural rigidity ratio of bolt-laminated timber beams.

Specimen	M_y (kN.mm)	$f_{b\text{-lam}}$ (MPa)	f_b (MPa)	k_S
BLB-01	3000.478	11.72	37.96	0.31
BLB-02	3268.675	12.77	37.96	0.34
BLB-03	3034.425	11.85	37.96	0.31
Average	3101.193	12.11	37.96	0.32

The bending strength of bolt-laminated timber beams obtained from an experimental test is 12.11 MPa (average) in the term of proportional limit load, while the bending strength of a solid timber beam is 37.96 MPa (secondary data from Atlas elastic cross-section modulus, timber, beam, bolt-laminated, Jabon Indonesia). The results showed that the elastic cross-section modulus and the flexural stress of the bolt-laminated timber beams were lower with a ratio of 0.32 than the solid timber beam, and the ductility ratio of the bolt-laminated timber beams was 1.18 so that they were categorized in the limited ductility criteria. The test results indicate that the failure of bolt-laminated Jabon timber beams are a failure of bending. The elastic modulus cross-section parameters of laminated timber beams are useful for the design of beam structure components in buildings, especially in the calculation of beam strength and beam stiffness as a serviceability requirement.

4. Conclusion

The load-deflection curve of bolt-laminated timber beams shows a bilinear trend. The presence of a bolt gives the beam a ductile impact. Changes in the behavior of beams from elastic to plastic conditions are due to the effective stress that occurs in some parts of the timber material that has exceeded the yield limit criteria. When the proportional limit load is reached, the bolt has not yielded, so that after the beam has post-elastic behavior the bolt still functions as a shear transfer between laminae, likewise the normal stress that occurs on each of the outer edge fibres has not exceeded the ultimate limit criteria so that the beam is still capable, endure the bending moment. The failure behavior of bolted-laminated timber beam occurs due to bending failure.

References

[1] Kramer V, Blass HJ, 2001, Load carrying capacity of nail-laminated timber under concentrated loads, Proceedings of 34th CIB W18 Meeting, Venice, Italy, CIB W18/34-12-5.

[2] American Society for Testing and Materials 2014 ASTM D143-14 Standard Test Methods for Small Clear Specimens of Timber, ASTM International, West Conshohocken, PA.

[3] Pranata Y A, Suryoatmono B, Tjondro J A 2012 Rasio Modulus Penampang Elastik Balok Kayu Laminasi-Baut, Jurnal Teknik Sipil, Volume 19 Nomor 3, Desember 2012, pp 223-236

[4] Pranata Y A, Suryoatmono B, Tjondro J A 2011 The Flexural Rigidity Ratio of Indonesian Timber Bolt-Laminated Beam, Prosiding The 3rd European Asian Civil Engineering Forum, Universitas Atma Jaya Yogyakarta, Indonesia.

[5] Pusat Penelitian dan Pengembangan Keteknikan Kehutanan dan Pengolahan 2013 Atlas Kayu \ Indonesia, Badan Penelitian dan Pengembangan Kehutanan, Kementerian Kehutanan

[6] Gere J M, Goodno B J 2018 Mechanics of Materials (Eighth Edition), ISBN: 978-1-337-09334-7, Cengage Learning, 20 Channel Center Street, Boston, MA 02210, USA

[7] Munoz W, Mohammad M, Salenikovich A, Quenneville P, 2010, Determination of Yield Point \ and Ductility of Timber Assemblies: In Search for a Harmonized Approach, Engineered Wood Products Association.

The Effect of E-WOM, Brand Trust, and Brand Ambassador on Purchase Decisions at Tokopedia Online Shopping Site

Nurhasanah[1*], Febrina Mahliza[1], Lucky Nugroho[1] and Yananto Mihadi Putra[1]
[1]Faculty of Economics and Business, Universitas Mercu Buana, Meruya Selatan Street, No 1, Kembangan, Jakarta Barat, Indonesia

E-mail: nurhasanah1619@gmail.com

Abstract. Online shopping sites are places for buying and selling transactions that are favored by many people. Tokopedia is currently the market leader in the world of e-commerce in Indonesia. The purpose of this study was to analyze the influence of electronic word of mouth (E-WOM), brand trust, and brand ambassador toward purchase decisions at the Tokopedia online shopping site. This study uses a quantitative approach by collecting data through the survey method. The population in this study were Tokopedia users from Mercu Buana University students. The data of 100 respondents were collected using an online questionnaire and were analyzed using Partial Least Square (PLS). This study shows that e-WOM has a positive and significant effect on purchase decisions. Brand trust has a positive and significant effect on purchase decisions, and brand ambassador has a positive and significant effect on purchase decisions.

1. Introduction

Internet users in Indonesia in 2017 were 143.26 million (54%) from a total of 264 million people [1]. The number of internet users has positive implications for the increasing intensity of online shopping in Indonesia. As many as 2.43 million people from the total internet users use the internet to shop online, while 2.72 million people also use the internet to sell online. This shows that 5.15 million people in Indonesia use the internet as a medium for buying and selling online. Technology, especially information technology, has a significant impact on various aspects, including economic and cultural life [2]. The rapid development of information technology influences the speed and convenience of services to customers [3]. In this condition, competition in the world of online sales is getting tougher. The development of buying and selling transactions in the digital world is also multiplying, which shows that people's purchasing decisions for shopping at online stores are also increasing every year. These online transactions are from big cities and small and rural towns in Indonesia. According to 2018, Tokopedia's transaction value was the US $ 5.9 billion [4]. This data shows that Tokopedia is currently the market leader in the world of e-commerce in Indonesia. Various factors can influence purchasing decisions at online shopping sites. Online purchases are strongly affected by E-WOM compared to other traditional communication media such as advertisements and the purchase of editorial recommendations. Reviews of a product made by consumers on the internet are a basic form of E-WOM. Consumers tend to look for product reviews online to obtain specific product information in the purchasing decision process. E-WOM has a positive and significant influence on purchasing decisions. As a review of a product or service in cyberspace, E-WOM is an important thing that can be used to increase someone's decision to buy a product or service. Online shopping sites that consumers see as a trustworthy brand can potentially determine consumer purchasing decisions. Consumers' feelings in trusting brand consumers can be remembered in consumers' minds, so consumers tend to choose the brand and will not switch to others. Brand trust has a positive and significant effect on purchase decisions [5]. The brand trust will determine consumer purchases of brands, and trust will potentially create high-value relationships. Besides, online shopping sites that use a brand ambassador can have the potential to influence consumers. It is intended that consumers are interested in making purchases, mostly since the selection of a brand ambassador is usually based on the image of a well-known celebrity. The brand ambassador has a positive and significant effect on purchasing decisions. This study is focused on providing empirical findings on online purchase decisions on an online shopping site in Indonesia that supports Micro, Small, and Medium Enterprises (MSMEs) and individuals to develop their businesses by marketing products online. Micro, Small, and Medium Enterprises (MSMEs) need to use information technology that plays a crucial role in improving their business performance.

2. Conceptual framework and hypothesis development

Consumer behavior is the study of how individuals, groups, and organizations select, buy, use, and dispose of goods, services, ideas, or experiences to satisfy their needs and wants [6]. It means that purchasing decisions show how individuals, groups, and organizations choose, buy, use, and how goods, services, ideas, or experiences satisfy their needs and wants. Purchasing decisions are preceded by the desire to buy, which arises because there are varying influencing factors such as family, the price desired, the information provided, and the benefits or benefits obtained from the product or service. When consumers make a purchase, many situational factors can influence their purchase decision. Consumers will undoubtedly look for information about the product or service they purchased before consumers purchase a product or service. Consumers' information about product quality and service quality needs crucial through E-WOM messages [7]. E-WOM is word of mouth through electronic media. E-WOM is a positive or negative statement made by potential consumers, real consumers, or consumers who have previously used a product about a product or company that can be accessed by many people or institutions via the internet. Consumers will undoubtedly look for information about the product or service they purchased before consumers purchase a product or service. E-WOM is the ability to drive the marketing of products issued by the company. E-WOM has a positive and significant influence on purchasing decisions [8]. The more positive the message conveyed via E-WOM, the higher the consumer's purchasing decision. This showed that E-WOM as a review of a product or service in cyberspace is an important thing that can increase someone's decision to buy a product or service. Therefore, the hypothesize that: H1: E-WOM has a positive and significant effect on purchase decisions at Tokopedia online shopping site.

Brand trust is a brand's ability to be trusted (brand reliability), which comes from consumer confidence that the product can fulfill the promised value and brand intention (brand intention), which is based on consumer confidence that the brand can prioritize. When a brand provides interaction to consumers, the brand can meet consumer needs and give responsibility for consumers' interests and safety. The impact is that consumers can feel very safe feeling towards the brand. Brand trust is consumers' feeling that consumers have trusted the product through product quality that can be remembered in consumers' minds. Furthermore, consumers tend to choose the product so that consumers will not switch to other products. The brand trust will determine consumer purchases of brands, and trust will potentially create high-value relationships [5] showed that brand trust has a positive and significant effect on purchasing decisions. The higher a person's perception of brand trust, the higher the purchase decision. Therefore, the hypothesize that H2: Brand trust has a positive and significant effect on purchase decisions at Tokopedia online shopping site.

A brand ambassador is a person who supports a brand from various famous public figures, apart from popular society can also be supported by ordinary people. Brand Ambassadors are cultural icons or identities, where they act as marketing tools that represent the achievement of human individualism and the commodification and commercialization of a product. The characteristics of the Brand

Ambassador are visibility, credibility, attraction, and power. Companies use Brand Ambassador to influence or engage consumers. It is intended that consumers are interested in making purchases, mainly because of the selection of a Brand Ambassador. Brand Ambassador has a positive and significant effect on purchasing decisions [9]. The higher a person's perception of a brand ambassador, the higher the purchase decision. Therefore, the hypothesize that: H3: Brand Ambassador has a positive and significant effect on purchase decisions at Tokopedia online shopping site.

3. Method

The type of study is causal research using a quantitative approach. This study's independent variables are E-WOM, brand ambassador, and brand trust. The dependent variable is the purchase decision, and the variables used as latent variables, which consist of indicators. The indicators of E-WOM are frequency of access to information from social networking sites, the number of reviews written by users of social networking sites, positive comments from users of social networking sites, recommendations from users of social networking sites, information on food and beverage variations, information on the quality (taste, texture, and temperature) of food and beverages and information regarding the price offered [10]. The population of this study was students of Mercu Buana University who had shopped at Tokopedia. The samples were determined by using the purposive sampling technique. The sample criteria are the students aged 20-32 years and had shopped at Tokopedia once (Table 1). The number of samples was 100 respondents, determined from 5 times ten indicators used in this study. SEM analysis relies on the number of parameters estimated where the sample is obtained from 5-10 times the number of parameters estimated [11].

Table 1. Profile of respondents.

Characteristics	Number	Percentage
Gender		
• Female	40	40%
• Male	60	60%
Age		
• 17-22	54	54%
• 23-27	34	34%
• 28-32	12	12%
Total	100	100%

This study was started from August 2019 to May 2020. The data collected using the distribution of online questionnaires to respondents. The data collection technique used in this research is distributing online questionnaires to respondents. The questionnaire has been tested for validity and reliability. The questionnaire is distributed using a Likert scale of 1 (strongly disagree) to 5 (strongly agree). Data were analyzed using partial least square (PLS). PLS consists of testing the measurement model used to test validity and reliability, while the structural model is used for the causality test [12]. PLS is an analysis method that is soft modeling because it does not assume data with a specific scale measurement, which means that the number of samples can be small.

4. Results and discussion

4.1. Results

4.1.1. Results of testing the outer model.

The convergent validity testing results on all indicators have met the convergent validity because of all loading factor values above 0.60. So, it reflects that all indicators are valid. Besides, the values of AVE must be above 0.5. From the results, all variables are valid (Table 2).

Table 2. Outer loading test results.

Variable	Indicators	Outer Loading	AVE	Composite Reliability	Cronbach's Alpha
Purchase Decisions	KP1	0,868	0.705	0.934	0.901
	KP2	0,923			
	KP3	0,873			
	KP4	0,824			
	KP5	0,815			
	KP6	0,720			
E-WOM	EW1	0,891	0.942	0.988	0.842
	EW2	0,790			
	EW3	0,871			
	EW4	0,906			
	EW5	0,807			
	EW6	0,668			
Brand Trust	BT1	0,879	0.769	0.909	0.847
	BT2	0,936			
	BT3	0,813			
Brand Ambassador	BA1	0,964	0.579	0.892	0.985
	BA2	0,966			
	BA3	0,985			
	BA4	0,968			
	BA5	0,970			

The next examination is discriminant validity. A construct with good discriminant validity if each loading factor value of each indicator of a latent variable has the most significant loading factor value with other loading values against other latent variables (Table 3).

Table 3. Results of discriminant validity test (cross-loading).

Indicators	Brand Ambassador	Brand Trust	E-WOM	Purchase Decision
BA1	**0.964**	0.480	0.609	0.647
BA2	**0.966**	0.545	0.681	0.720
BA3	**0.985**	0.524	0.664	0.681
BA4	**0.968**	0.532	0.714	0.733
BA5	**0.970**	0.516	0.668	0.671
BT1	0.537	**0.879**	0.636	0.758
BT2	0.492	**0.936**	0.749	0.815
BT3	0.364	**0.813**	0.575	0.601
EW1	0.248	0.509	**0.750**	0.465
EW2	0.488	0.596	**0.801**	0.666
EW3	0.292	0.469	**0.752**	0.474
EW4	0.284	0.537	**0.780**	0.498
EW5	0.619	0.699	**0.807**	0.923
EW6	0.970	0.516	**0.668**	0.671
KP1	0.545	0.665	0.701	**0.868**
KP2	0.619	0.699	0.807	**0.923**
KP3	0.485	0.695	0.744	**0.873**
KP4	0.465	0.609	0.637	**0.824**
KP5	0.492	0.936	0.749	**0.815**
KP6	0.966	0.545	0.681	**0.720**

This can be seen in the indicators of E-WOM, brand trust, brand ambassador, and purchase decisions that latent constructs predict indicators in their block better than indicators in other blocks. Thus, it can be concluded that from the results of the cross-loading analysis, it appears that there is no discriminant validity problem. The results of testing for composite reliability and Cronbach alpha also show a fair value, that is, all latent variables are reliable because all latent variable values have a composite reliability value and Cronbach's alpha ≥ 0.70 (Table 2). So it can be concluded that the variables used have been consistent.

4.1.2. Results of testing the inner model.
In hypothesis testing, the E-WOM variable has a positive and significant effect on purchasing decisions (t-statistics 5.116 is greater than the t-table value of 1.6581, and the coefficient value is 0.406). Brand Trust has a positive and significant effect on purchasing decisions (t-statistics 4,616 is greater than the t-table value of 1.6581, and the coefficient value is 0.417). Brand Ambassador has a positive and significant effect on purchasing decisions (t-statistics 3.244 greater than the t-table value of 1.6581 and the coefficient value of 0.209). Thus the three hypotheses are accepted.

Table 4. Hypothesis testing results.

Path	Original Sample	T-Statistics	Results
E-WOM → Purchase Decisions	0.406	5.116	Accepted
Brand Trust → Purchase Decisions	0.417	4.616	Accepted
Brand Ambassador → Purchase Decisions	0.209	3.244	Accepted

The R-square value in this study was 0.848. This means that the three variables can explain the variability of purchasing decisions in the model, namely E-WOM, brand trust, and the brand ambassador is 84.8%. Meanwhile, 15.2% is explained by other variables not examined in this model.

4.2. Discussion

4.2.1. The effect of E-WOM on purchase decisions.
Based on the results, E-WOM has a positive and significant effect on purchase decisions, where the better product reviews are, the more people decide to buy products at Tokopedia. These results show similarities with research conducted by [13], which reveal that electronic word of mouth has a positive effect on purchase decisions. The E-WOM variable in this study is reflected through the frequency of access to information from social networking sites, the number of reviews written by users of social networking sites, positive comments from users of social networking sites, recommendations from users of social networking sites, information on food and beverage variations, information on the quality (taste, texture, and temperature) of food and beverages and information regarding the price offered. Social networking sites' recommendations become the most dominant contributing indicator in reflecting the E-WOM variable. Respondents considered that there are recommendations from users of social networking sites to shop at Tokopedia. It is essential as an indicator of electronic word of mouth, which in turn has an impact on influencing consumers to buy at Tokopedia. Many companies rely on information nowadays. E-WOM is useful because communication between people is related to excellence or experience when buying or using a product or service to make purchase decisions.

4.2.2. The effect of brand trust on purchase decisions.
Based on the results, brand trust has a positive and significant effect on purchase decisions. The higher the trust in the Tokopedia Brand (brand trust), the higher the people will decide to buy products atTokopedia. These results are supported by research conducted by [14] that brand trust positively affects purchasing decisions. The belief that brands provide a sense of security becomes the most dominant contributing indicator in reflecting the brand trust variable. Respondents considered that

Tokopedia always fulfills its promises to consumers, such as improving security on Tokopedia. Before a consumer makes online purchasing, they trust brands that provide security [15]. It is essential as an indicator of trusting Tokopedia, which in turn has an impact on influencing consumers to buy at Tokopedia. Brand Trust is the variable that has the most significant influence on purchasing decisions. Brand Trust is built because of the trust and hope that Tokopedia will act according to consumers' needs and desires. A company (brand) with satisfactory performance will be considered to have good value [16]. When someone has trusted the other party, they believe that expectations will be fulfilled and will not be disappointed. This must be considered so that consumers can more determine the right purchase decision.

4.2.3. The effect of brand ambassador on purchasing decisions.
Based on the results, the brand ambassador has a positive and significant effect on purchase decisions. The more attractive the brand ambassador is, the more people decide to buy products at Tokopedia. This result is supported by research conducted by [17], which revealed that brand ambassadors positively affect purchase decisions. This study's brand ambassador variable is reflected through transference, congruence, credibility, attractiveness, and power. Credibility has become the most dominant contributing indicator in reflecting the brand ambassador variable. Respondents considered that Tokopedia uses a brand ambassador that has high credibility and can increase sales on Tokopedia. This is also proven by using BTS to become Tokopedia's Brand Ambassador. It is essential as an indicator of brand ambassador, which in turn has an impact on influencing consumers to buy at Tokopedia. All business actions must change to follow the market and environmental developments [18]. Brand ambassadors drive brand awareness and attract new customers to make purchasing decisions.

5. Conclusion
This study is intended to analyze whether E-WOM, brand trust, and brand ambassador affect purchasing decisions on Tokopedia. E-WOM has a positive and significant effect on purchasing decisions, which means that more and better product reviews will make people decide to buy products on Tokopedia. Brand trust has a positive and significant effect on purchase decisions, which means that online shopping site providers with high brand trust will influence someone to make online purchasing. The brand ambassador has a positive and significant effect on purchase decisions, which means that the more credible a brand ambassador is, the higher the purchase decision of a consumer on an online shopping site.

References
[1] Indonesia Internet Service Provider Association. Infografis Penetrasi & Profil Perilaku Pengguna Internet Indonesia. 2017.
[2] Putra YM. Analysis of Factors Affecting the Interests of SMEs Using Accounting Applications. J Econ Bus. 2019;2(3).
[3] Nugroho L, Chowdhury SLK. Mobile Banking for Empowerment Muslim Women Entrepreneur: Evidence from Asia (Indonesia and Bangladesh). Tazkia Islam Financ Bus Rev [Internet]. 2015 Jan 4;9(1):83–100. Available from: http://tifbr-tazkia.org/index.php/TIFBR/article/view/79
[4] CLSA Indonesia. Tokopedia, E-Commerce dengan Nilai Transaksi Terbesar. 2019.
[5] Wulandari NMR, Nurcahya IK. Pengaruh Celebrity Endorser, Brand Image, Brand Trust Terhadap Keputusan Pembelian Clear Shampoo Di Kota Denpasar. E-Jurnal Manaj Univ Udayana. 2015;4(11).
[6] Kotler P. Reconceptualizing marketing: An interview with Philip Kotler. Eur Manag J. 1994 Dec 1;12(4):353–61.
[7] Chevalier JA, Mayzlin D. the Effect of Word of Mouth on Sales: Online Book Reviews. NBER Working Paper Series. 2003. (No. 10148).

[8] Cheung CMK, Lee MKO, Thadani DR. The impact of positive electronic word-of-mouth on consumer online purchasing decision. In: In World Summit on Knowledge Society. 2009. p. 501–10.
[9] Nugroho L. Pengetahuan Konsumen Vis A Vis Perubahan Perilaku Konsumen Pada Era Digital. In: Perilaku Konsumen dan Perkembangannya di Era Digital. Widina Bhakti Persada; 2020. p. 65–77.
[10] Priansa DJ. Pengaruh E-WOM dan Persepsi Nilai terhadap Keputusan Konsumen untuk Berbelanja Online di Lazada. Ecodemica. 2016;IV(1):117–24.
[11] Hair JF, Black WC, Babin BJ, Anderson RE. Multivariate Data Analysis. 7th ed. New Jersey: Prentice Hall; 2010.
[12] Abdillah W, Jogiyanto HM. Partial Least Square (PLS) Alternatif Structural Equation Modeling (SEM) dalam Penelitian Bisnis. Yogyakarta: Andi; 2015.
[13] Faza MA. Pengaruh Electronic Word of Mouth Terhadap Minat Beli Melalui Citra Merek (Studi Pada Smartphone Xiaomi Di Kota Yogyakarta). J Fak Ekon. 2018;6(4):1–11.
[14] Antoni, Herianto M. Pengaruh Brand Image dan Kepercayaan Merek terhadap Minat Beli Konsumen pada Produk Mobil Toyota Etios Valco (Studi pada PT. Agung Automall Pekanbaru). JOM FISIP. 2017;4(1):1–10.
[15] Mahliza F. Consumer Trust in Online Purchase Decision. EPRA Int J Multidiscip Res (IJMR)-Peer Rev J. 2020;6(2):142–9.
[16] Anggraini D, Tanjung PRS. Company Value: Disclosure Implications of Sustainable Supply Chian, Profitability and Industrial Profile. Int J Supply Chain Manag. 2020;9(2):648–55.
[17] Samosir LS, Putri YR, Nurfebrianing S. Pengaruh Penggunaan Brand Ambassador Dewi Sandra Terhadap Putusan Pembelian Kosmetik Wardah Di Kota Bandung. J Sosioteknologi. 2016;15(2):233–40.
[18] Rekarti E, Doktoralina CM. Improving Business Performance: A Proposed Model for SMEs. Eur Res Stud J. 2017;20(3):613–23.

IOP Conf. Series: Materials Science and Engineering 1071 (2021) 012018 doi:10.1088/1757-899X/1071/1/012018

Mapping the Spread of Covid-19 in Asia Using Data Mining X-Means Algorithms

N Arminarahmah[1], A Daengs GS[2], G W Bhawika[3], M P Dewi[4] and A Wanto[5*]
[1] Universitas Islam Kalimantan Muhammad Arsyad Al Banjari, Banjarmasin, Indonesia
[2] Universitas 45 Surabaya, Surabaya, Indonesia
[3] Institut Teknologi Sepuluh Nopember, Surabaya, Indonesia
[4] Institut Teknologi dan Bisnis Asia Malang, Malang, Indonesia
[5] STIKOM Tunas Bangsa, Medan, Indonesia

E-mail: anjarwanto@amiktunasbangsa.ac.id

Abstract. Until now, Covid-19 is a phenomenal problem faced by almost all over the world, especially countries on the Asian continent. Not only causing casualties, but this virus also affects the wheels of the country's economy. The purpose of this paper is to view and map the spread of the Covid-19 virus in Asia based on Total Cases, Total Deaths, Total Cured and Active Cases from 49 Countries. The research data in this paper were obtained from the Worldometer website sourced from WHO, CDC, NHC and others. In this proposed paper, the algorithm used is X-Means Clustering with the help of Rapidminer. The results of this proposed paper are in the form of grouping or mapping the spread of the Covid-19 virus in Asia which is divided into 4 zones, including the Red Zone (the number of active cases of Covid -19 and the death rate is very high) which consists of 1 country, the Orange Zone (number of cases active covid -19 and the mortality rate is quite high) consisting of 1 countries, the Yellow Zone (active case rate of Covid -19 and moderate mortality) consisting of 39 countries, and the Blue Zone (active case rate of covid -19 and low mortality rate) which consists of 8 countries.

1. Introduction

The coronavirus pandemic until the end of 2020 is still a phenomenal problem faced by almost all over the world since it was first identified in Wuhan last December 2019 and it is not certain exactly when it will end. This includes countries in the Asian continent, which is the largest continent in the world. As previously known, the COVID-19 disease, which is caused by the severe acute respiratory syndrome coronavirus 2 (SARS-CoV-2), is a major global crisis [1][2]. On March 11, 2020, the World Health Organization (WHO) declared COVID-19 a pandemic, and it has become one of the deadliest pandemics in the past century [3]. Not only causing casualties, but this virus also affects the wheels of the country's economy. This paper will discuss the mapping of the distribution of Covid-19 in Asian countries, which consist of 49 countries, this is because this virus was first identified in China, which is one of the countries on the Asian continent. Based on data obtained from the Worldometers website sourced from WHO, CDC and NHC, the number of Covid-19 cases in the world as of Sunday 13 December 2020 at 11.22 WIB reached 72,653,057 people with a total death reaching 1,618,890

IOP Conf. Series: Materials Science and Engineering 1071 (2021) 012018 doi:10.1088/1757-899X/1071/1/012018

people, while in Asia from In 49 countries, the number of Covid-19 cases reached 19,084,461 people with a total death toll of 312,236 [4].

Therefore, mapping the distribution of Covid-19 is very important to do, as information and a barometer for governments in Asian countries, especially for the Indonesian government in determining policies related to this matter. The specific purpose of this research is to provide input and information for the Indonesian government to maximize efforts to be able to overcome the Covid-19 problem and for the Indonesian people to temporarily not travel to countries in Asia that are included in the Red, Orange and Yellow Zones. The research dataset is a dataset of Covid-19 cases in Asia from the Worldometers website sourced from WHO, CDC and NHC. The mapping algorithm used in this study is the X-Means Algorithm.

Table 1. Data on Covid-19 Cases in Asia

No	Country	Total Cases	Total Deaths	Total Recovered	Active Cases
1	Afghanistan	49.273	1.971	38.384	8.918
2	Armenia	148.325	2.486	125.169	20.670
3	Azerbaijan	175.874	1.922	111.918	62.034
4	Bahrain	89.143	348	87.182	1.613
5	Bangladesh	490.533	7.052	420.896	62.585
6	Bhutan	437	0	400	37
7	Brunei	152	3	147	2
8	Cambodia	359	0	307	52
9	China	86.725	4.634	81.785	306
10	Cyprus	15.101	78	2.057	12.966
11	Georgia	189.726	1.790	158.183	29.753
12	Hong Kong	7.542	117	6.202	1.223
13	India	9.884.716	143.393	9.387.608	353.715
14	Indonesia	617.820	18.819	505.836	93.165
15	Iran	1.108.269	52.196	812.270	243.803
16	Iraq	574.634	12.579	507.446	54.609
17	Israel	357.176	2.999	337.186	16.991
18	Japan	177.287	2.562	149.286	25.439
19	Jordan	259.614	3.365	218.467	37.782
20	Kazakhstan	141.578	2.088	125.380	14.110
21	Kuwait	146.218	911	142.094	3.213
22	Kyrgyzstan	77.356	1.307	70.192	5.857
23	Laos	41	0	33	8
24	Lebanon	146.520	1.200	101.334	43.986
25	Macao	46	0	46	0
26	Malaysia	83.475	415	69.393	13.667
27	Maldives	13.368	48	12.706	614
28	Mongolia	907	0	384	523
29	Myanmar	108.342	2.268	86.795	19.279
30	Nepal	248.423	1.698	235.731	10.994
31	Oman	126.240	1.471	118.048	6.721
32	Pakistan	438.425	8.796	383.000	46.629
33	Palestine	109.738	978	85.110	23.650
34	Philippines	449.394	8.733	418.682	21.979
35	Qatar	140.961	240	138.490	2.231
36	S. Korea	42.766	580	31.814	10.372
37	Saudi Arabia	359.888	6.048	350.549	3.291
38	Singapore	58.320	29	58.208	83
39	Sri Lanka	32.790	152	23.793	8.845
40	Syria	9.166	518	4.376	4.272
41	Taiwan	736	7	606	123

No	Country	Total Cases	Total Deaths	Total Recovered	Active Cases
42	Tajikistan	12.704	88	12.133	483
43	Thailand	4.209	60	3.923	226
44	Timor-Leste	31	0	30	1
45	Turkey	1.836.728	16.417	1.603.780	216.531
46	UAE	184.949	617	163.679	20.653
47	Uzbekistan	74.956	612	72.243	2.101
48	Vietnam	1.397	35	1.241	121
49	Yemen	2.083	606	1.383	94
	Total	**19.084.461**	**312.236**	**17.265.905**	**1.506.320**

X-Means is a data mining algorithm which is a development of the K-Means, and this algorithm can cover the shortcomings of the K-Means [5]. As is well known, data mining algorithms are algorithms that are often used for data classification [6]–[10], data clustering [11]–[14] and Prediction [15]–[18].

Several previous studies related to the topic of this article include: Geo-Social Big Data Clustering using Focused Information Criterion based Partitioned Iterative X-Means Dice Correlation Data Clustering (FIC-PIXDCDC) Method. In this algorithm, geosocial data taken as input (user, location, and time) is obtained from the Weeplaces dataset. After that, this algorithm selects the number of clusters and centroids randomly. Then calculate the Dice Correlation between each geo-social data input and the cluster centroid. Furthermore, this algorithm applies Focused Information Criteria to build the optimal number of clusters for a certain large dataset. The results of this study indicate that this algorithm has higher accuracy and shorter time than other conventional algorithms [19]. Furthermore, the use of the Euclidean Distance X-Means algorithm for the BigData cluster. This algorithm is used to test data based on the centroid value which is then modified to obtain five clusters, based on the Euclidean distance. After testing, the data obtained from the cluster members have a similarity level to other data. The number of clusters determined by modification of the euclidean distance resulted in a better similarity level than if it was done randomly from each member [20]. The next research was to determine the X-Means centroid clustering using the Davies-Bouldin method. In this paper the X-Means algorithm is modified to perform several centroid determinations. So as to produce a better level of similarity of cluster members compared to other data. The Davies-Bouldin Index method is used to determine the centroid for testing two clusters with a minimum value with a DBI value close to 0 [21]. These related studies are behind the conduct of research to map the distribution of Covid-19 in Asia.

2. Methodology

2.1. Algorithms and Research Data
This study uses the X-Means Algorithm which is one of the Data Mining algorithms used to cluster data, while the research data is data on Covid-19 Cases in Asia obtained from the Worldometers website sourced from WHO, CDC and NHC (can be seen in Table 1).

2.2. X-Means Stages
The X-Means clustering stages in this study are presented in Figure 1.

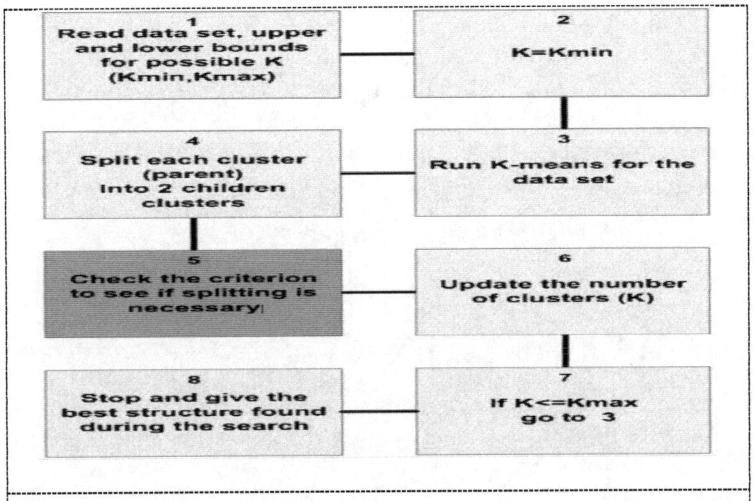

Figure 1. X-Means Stages

Defined K_{min} and $K_{ma}x$ as the lower and upper limits for possible K value K. In the initial X-Means clustering step, it is known that currently $K = K_{min}$, Initial and structures centroid are found by K-Means. In the next stage, each cluster in the presumed structure is treated as the parent cluster, which can be broken down into two child clusters. Based on several criteria, which will be described in the next section, we assessed the structure of the parent and child. Scores help to decide whether parents are a better representation of the sample data or clusters of children provide a more accurate distribution of the sample. As a result, the parent centroid will be replaced by the child centroid, or the algorithm will retain the parent centroid and leave the children. Then, new structures will be built or updated based on the choice of the parent or child. This procedure will continue for all clusters in the initial structure until the current estimated number of clusters is greater than K_{max} or the algorithm converges to the best structure.

3. Results and Discussion

3.1. Cluster Determination (K_{min} and K_{max})

In implementing the X-Means algorithm using Rapidminer, first is to determine the K_{min} value (the minimum number of clusters that must be detected) and the K_{max} value (the maximum number of clusters that must be detected). The value of Kmin and K_{max} can be determined randomly or randomly which is optional. There are 4 clusters defined in this paper.

3.2. Process X-Means with Rapidminer

The process of mapping the distribution of Covid-19 in Asia in the form of clustering and the results of the X-Means algorithm using Rapidminer can be seen in figure 2, figure 3, figure 4, figure 5 and figure 6.

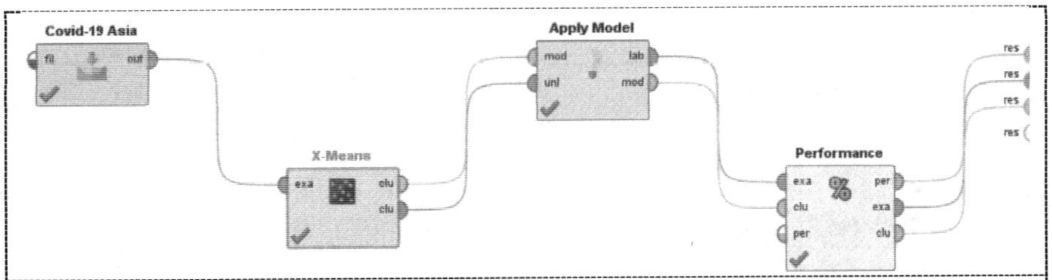

Figure 2. Process X-Means (K_{min} = 4 and K_{max} = 100)

Figure 2 describes the X-Means clustering process with Rapidminer which begins by entering Covid-19 data in Asia, followed by the selection of the X-Means operator with a value of $K_{min} = 4$ and a value of $K_{max} = 100$. Then it is connected to the Apply Model to apply a model that has been trained or studied. Then connect to Cluster Distance Performance to evaluate the performance of the model which provides a list of performance criteria values automatically according to the given task.

Cluster Model

```
Cluster 0: 39 items
Cluster 1: 1 items
Cluster 2: 1 items
Cluster 3: 8 items
Total number of items: 49
```

Figure 3. Cluster (X-Means)

Figure 3 is a cluster model produced by Rapidminer, Cluster_0: 39 items, Cluster_ 1: 1 item, Cluster_2: 1 item, and Cluster_3: 8 items. For the final result, the Centroid table can be seen in Figure 4.

Attribute ↑	cluster_0	cluster_1	cluster_2	cluster_3
Active Cases	10077.487	353715	216531	67881.500
Total Cases	76073.795	9884716	1836728	549517.375
Total Deaths	902.667	143393	16417	14652.750
Total Recovered	65093.641	9387608	1603780	466983.125

Figure 4. Centroid Table

The detailed results of mapping the distribution of Covid-19 in Asia can be seen in Figure 5.

- root
  - cluster_0
    - Afghanistan
    - Armenia
    - Azerbaijan
    - Bahrain
    - Bhutan
    - Brunei
    - Cambodia
    - China
    - Cyprus
    - Georgia
    - Hong Kong
    - Japan
    - Jordan
    - Kazakhstan
    - Kuwait
    - Kyrgyzstan
    - Laos
    - Lebanon
    - Macao
    - Malaysia
    - Maldives
    - Mongolia
    - Myanmar
    - Nepal
    - Oman
    - Palestine
    - Qatar
    - S. Korea
    - Singapore
    - Sri Lanka
    - Syria
    - Taiwan
    - Tajikistan
    - Thailand
    - Timor-Leste
    - UAE
    - Uzbekistan
    - Vietnam
    - Yemen
  - cluster_1
    - India
  - cluster_2
    - Turkey
  - cluster_3
    - Bangladesh
    - Indonesia
    - Iran
    - Iraq
    - Israel
    - Pakistan
    - Philippines
    - Saudi Arabia

Figure 5. Folder View Covid-19 Mapping and Spread in Asia

Cluster_0 based on figure 5 consists of 39 countries (Afghanistan to Yemen). Cluster_1 consists of 1 Country (India). Cluster 2 consists of 1 country (Turkey). Cluster_3 consists of 8 countries (Bangladesh, Indonesia, Iran, Iraq, Israel, Pakistan, Philippines and Saudi Arabia).

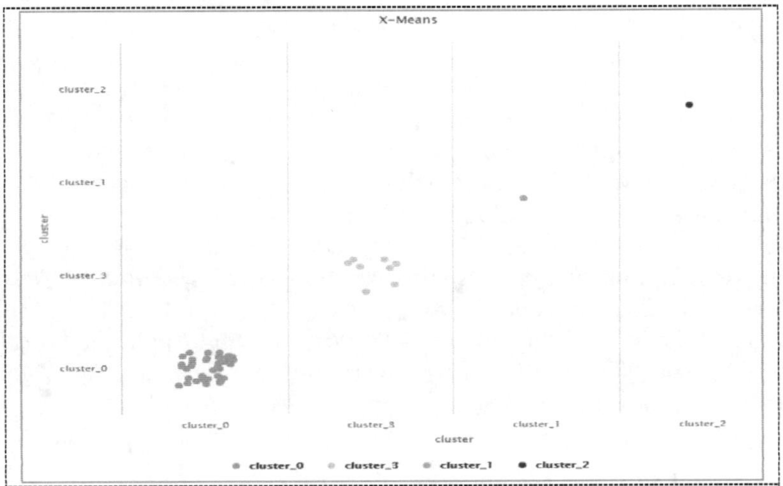

Figure 6. Mapping the Covid-19 Spread in Asia

The blue dot in the figure is the Covid-19 Cluster_0 (Blue Zone) Spread Mapping. The green dot represents the Covid-19 Cluster_3 Spread Mapping (Yellow Zone). The orange dot represents the Covid-19 Cluster_1 (Red Zone) Spread Mapping. The black dot represents the Covid-19 Cluster_2 Spread Mapping (Orange Zone).

4. Conclusion

Data Mining X-Means can be used to map the Covid-19 spread in Asia. Based on the results of the Covid-19 spread mapping analysis using Rapidminer, it is divided into 4 zones, namely the Red Zone (the number of active cases of Covid -19 and the death rate is very high) consisting of 1 country, namely India (Cluster_1), the Orange Zone (Number of active cases of Covid- 19 and the mortality rate is quite high) consists of 1 country, namely Turkey (Cluster-2), the Yellow Zone (the number of active cases of Covid -19 and moderate deaths) consists of 8 countries, including: Bangladesh, Indonesia, Iran, Iraq, Israel, Pakistan , Philippines and Saudi Arabia (Cluster_3), and the Blue Zone (the number of active cases of Covid -19 and the low mortality rate) which consists of 39 countries, including Afghanistan to Yemen (Cluster_0), as presented in figure 5. Performance Vector X-Means in this paper results in the value Avg._within_centroid_distance = -20930456418.490 and Davies_Bouldin = -0.266.

References

[1] Z. J. Cheng and J. Shan, "2019 Novel coronavirus: where we are and what we know," *Infection*, vol. 48, no. 2, pp. 155–163, 2020.

[2] F. He, Y. Deng, and W. Li, "Coronavirus disease 2019: What we know?," *Journal of Medical Virology*, vol. 92, no. 7, pp. 719–725, 2020.

[3] M. R. Azarpazhooh *et al.*, "COVID-19 pandemic and burden of non-communicable diseases: An ecological study on data of 185 countries," *Journal of Stroke and Cerebrovascular Diseases*, vol. 29, no. 9, pp. 1–9, 2020.

[4] Worldometer, "Reported Cases and Deaths by Country, Territory, or Conveyance," *COVID-19 Coronavirus Pandemic*, 2020. [Online]. Available: https://www.worldometers.info/coronavirus/. [Accessed: 15-Dec-2020].

[5] A. Radwan *et al.*, "X-means Clustering for Wireless Sensor Networks," *Journal of Robotics, Networking and Artificial Life*, vol. 7, no. 2, pp. 111–115, 2020.

[6] I. S. Damanik, A. P. Windarto, A. Wanto, Poningsih, S. R. Andani, and W. Saputra, "Decision Tree Optimization in C4.5 Algorithm Using Genetic Algorithm," *Journal of Physics: Conference Series*, vol. 1255, no. 1, pp. 1–6, Aug. 2019.

[7] W. Katrina, H. J. Damanik, F. Parhusip, D. Hartama, A. P. Windarto, and A. Wanto, "C.45 Classification Rules Model for Determining Students Level of Understanding of the Subject," *Journal of Physics: Conference Series*, vol. 1255, no. 1, pp. 1–7, 2019.

[8] H. Siahaan, H. Mawengkang, S. Efendi, A. Wanto, and A. Perdana Windarto, "Application of Classification Method C4.5 on Selection of Exemplary Teachers," *Journal of Physics: Conference Series*, vol. 1235, no. 1, 2019.

[9] I. Parlina *et al.*, "Naive Bayes Algorithm Analysis to Determine the Percentage Level of visitors the Most Dominant Zoo Visit by Age Category," *Journal of Physics: Conference Series*, vol. 1255, no. 1, pp. 1–5, 2019.

[10] D. Hartama, A. Perdana Windarto, and A. Wanto, "The Application of Data Mining in Determining Patterns of Interest of High School Graduates," *Journal of Physics: Conference Series*, vol. 1339, no. 1, pp. 1–6, 2019.

[11] M. A. Hanafiah, A. Wanto, and P. B. Indonesia, "Implementation of Data Mining Algorithms for Grouping Poverty Lines by District/City in North Sumatra," *International Journal of Information System & Technology*, vol. 3, no. 2, pp. 315–322, 2020.

[12] N. A. Febriyati, A. D. GS, and A. Wanto, "GRDP Growth Rate Clustering in Surabaya City uses the K- Means Algorithm," *International Journal of Information System & Technology*, vol. 3, no. 2, pp. 276–283, 2020.

[13] S. Sudirman, A. P. Windarto, and A. Wanto, "Data Mining Tools | RapidMiner : K-Means Method on Clustering of Rice Crops by Province as Efforts to Stabilize Food Crops In Indonesia," *IOP Conference Series: Materials Science and Engineering*, vol. 420, no. 012089, pp. 1–8, 2018.

[14] S. A. Abbas, A. Aslam, A. U. Rehman, W. A. Abbasi, S. Arif, and S. Z. H. Kazmi, "K-Means and K-Medoids: Cluster Analysis on Birth Data Collected in City Muzaffarabad, Kashmir," *IEEE Access*, vol. 8, pp. 151847–151855, 2020.

[15] M. Tsiakmaki, G. Kostopoulos, S. Kotsiantis, and O. Ragos, "Implementing AutoML in Educational Data Mining for Prediction Tasks," *Applied Sciences*, vol. 10, no. 90, pp. 1–27, 2020.

[16] A. K. Verma, S. Pal, and S. Kumar, "Prediction of Skin Disease Using Ensemble Data Mining Techniques and Feature Selection Method-a Comparative Study," *Applied Biochemistry and Biotechnology*, pp. 1–19, 2019.

[17] S. Borah, V. E. Balas, and Z. Polkowski, *Advances in Data Science and Management*. Springer, 2019.

[18] S. Q. Salih *et al.*, "River suspended sediment load prediction based on river discharge information: application of newly developed data mining models," *Hydrological Sciences Journal*, vol. 65, no. 4, pp. 624–637, 2020.

[19] M. Anoop and P. Sripriya, "Focused information criterion based partitioned iterative X-means dice correlation clustering for big Geo-social data," *Journal of Critical Reviews*, vol. 7, no. 6, pp. 54–62, 2020.

[20] N. Zendrato, H. W. Dhany, N. A. Siagian, and F. Izhari, "Bigdata Clustering using X-means method with Euclidean Distance," *Journal of Physics: Conference Series*, vol. 1566, no. 1, p. 012103, 2020.

[21] M. Mughnyanti, S. Efendi, and M. Zarlis, "Analysis of determining centroid clustering x-means algorithm with davies-bouldin index evaluation," *IOP Conference Series: Materials Science and Engineering*, vol. 725, no. 1, p. 012128, 2020.

IOP Conf. Series: Materials Science and Engineering 1071 (2021) 012019 doi:10.1088/1757-899X/1071/1/012019

Development of Virtual Reality Application to Increase Student Learning Motivation with Interactive Learning in Vocational Education

F I Maulana[1*] and A Purnomo[2]
[1]Computer Science Department, School of Computer Science, Bina Nusantara University, Indonesia
[2]Entrepreneurship Department, BINUS Business School Undergraduate Program, Bina Nusantara University Jakarta, Indonesia

E-mail: fairuz.maulana@binus.edu

Abstract. In Vocational Education, learning by doing is very important to complement the concept of learning with technical introduction to maximize the results of the material received. However, in various subject areas, learning is still carried out conventionally and students are unable to apply their knowledge practically even though they have theoretical understanding. This is closely related to the lack of motivation from students in the classroom because there is no interactive course content required for visualization of concepts in each subject area. Even during the COVID-19 pandemic, teaching and learning activities had to be carried out online at their respective homes. So that further limits the scope of student practice to maximize learning outcomes. As a result, it is not possible for students to understand more in-depth concepts of subject matter using conventional learning methods. Students' motivation to learn will increase if students are provided with practical tools so that they can be visually involved and engage with what they teach in class. The development of practical tools with Virtual Reality technology, which can visualize 3D models can be a solution to meet this goal. The main objective of this paper is to create and use Virtual Reality as an immersive learning medium and its contribution to student motivation to understand the definition of subject matter in various fields.

1. Introduction

A practical way of communicating ideas or concepts to others is through digital presentations that have utilized current information and communication technology, one of which is Virtual Reality technology [1]. Virtual reality is a system that allows users in a virtual universe that is replicated by a computer to communicate with an actual environment, so that users believe they are in that environment.

The Industrial Revolution 4.0 encourages the world of education to take part in the development of educational technology. In many laws and regulations of the Republic of Indonesia, the government has restricted the need for technology. One of them is specified in Permendiknas No. 41 of 2007 on Processing Requirements, in compliance with the Skill Standards for Graduates and Contents of the Learning Concepts used in item 13 of Permendiknas No. 16 of 2007 on the use of information and communication technologies to enhance the quality and effectiveness of learning.

With VR, students can do vocational training and enter the workplace environment according to their vocational, for example students from a culinary vocational can feel the atmosphere in a luxury restaurant kitchen, or students with medical majors who can experience situations facing critical patients, all of that becomes possible virtually. reality. This technology can increase efficiency in the classroom and provide students with hands-on experiences [2].

VR for education will also allow students to face problems in their own way, building knowledge from their experiences. So, students can engage in authentic problems, explore solutions and can collaborate with others. Remarkably, this method was successful in increasing the learning ability of students who performed quite low [3]. The potential that VR has for increasing the potential for learning and student engagement cannot be underestimated. In a world increasingly based on experience, the value of VR continues to grow [4].

The reality of the implementation that occurs in classrooms today, teachers still use instructional techniques and books as learning materials, which are conventional learning methods and have not been carried out effectively and interactively. This will have an impact on the decline in student interest in learning because students feel bored.

To increase students' interest in learning, the teacher's efforts are needed to design and implement a learning model that is appropriate, interactive, and fun [5]. Digital training materials need to be used in the Industrial era 4.0 to meet national competitiveness targets that would replace traditional learning models. Therefore, it is necessary to have learning media that can make students feel the process of the Paiton PLTU power plant from the coal carried by the barge to the occurrence of electrical energy. Industry 4.0 encourages the education revolution with immersive technology, one of which is Virtual Reality (VR).

2. Model Preparation and Design

This research was carried out by mapping starting from the scope of recommendations for Virtual Reality materials that adapt to the subject curriculum and the specifics of basic competencies in Vocational High Schools (SMK). The design of the Industry 4.0 Virtual Reality model activity was carried out in several sessions according to the designed timeline, as follows:

- Analysis:
  a. Analyzing the needs of stakeholders
  b. Identification of IQ & KD in the curriculum
  c. School Identification
- Design
- Development
- Evaluation & Dissemination

2.1. Analysis

First, assess the need for the instruction to be the answer. Conduct instructional research to assess target perceptual, affective, and motor skills targets. Determine what skills learners are supposed to have, as well as the effect of acquiring these skills. Analyze the time available and how it can be done over a period of time. Some scholars also recommend an overview of the context and available tools.

2.2. Design

Translating learning goals into cumulative success outcomes and primary objectives for each material unit. Determine the learning subjects or units to be discussed and how much time will be spent on each of them. The sequence of units relevant to the goals of the program. Refining the unit of training, defining the key tasks to be accomplished for each unit. Establish the lessons and learning tasks of each class. Establish specifications for determining what students are studying.

2.3. Development
Allow recommendations on the forms of learning experiences and instructional resources. Prepare draft materials and events. Try materials and events for target group members. Examination, repair and manufacture of goods and events. Produce teaching teachers or extra content.

2.4. Evaluation & Dissemination
Market materials to be used by future teachers and learners. Provide assistance or help if needed. Implement student appraisal plans. Implement software review plans. Establish strategies for the maintenance of learning and review tasks.

3. Result and Discussion

3.1. Formatting the title
Development of the Power Plant learning media using Virtual Reality based on the Vocational High School curriculum. The curriculum in question is Core Competencies (KI) and Basic Competencies (KD).

Table 1. Core Competencies and Basic Competencies.

CURRICULUM	
Core Competencies (KI)	***Basic Competencies (KD)***
Describe the function of the components in a power plant	Demonstrating the function of components in a power plant
Explain the diagram of the working principle of a Power Plant.	Demonstrating a diagram of the working principle of a Power Plant.

3.2. Asset Arrangement
Asset 3D models are made based on storyboards and discussions with teachers in schools as resource persons. Making 3D models of assets using Blender software and continued with Unity3D software to become a Power Plant VR application. The arrangement and placement of assets are adjusted to approach the existing Power Plant in the real world, namely the Power Plant Paiton East Java.

Figure 1. Paiton PLTU 3D Model

IOP Conf. Series: Materials Science and Engineering 1071 (2021) 012019 doi:10.1088/1757-899X/1071/1/012019

Figure 2. Paiton PLTU 3D Model

The next move is to assess the effectiveness of the product produced from the results of the development carried out, i.e. the "Power Plant" virtual reality educational media as in the following figure:

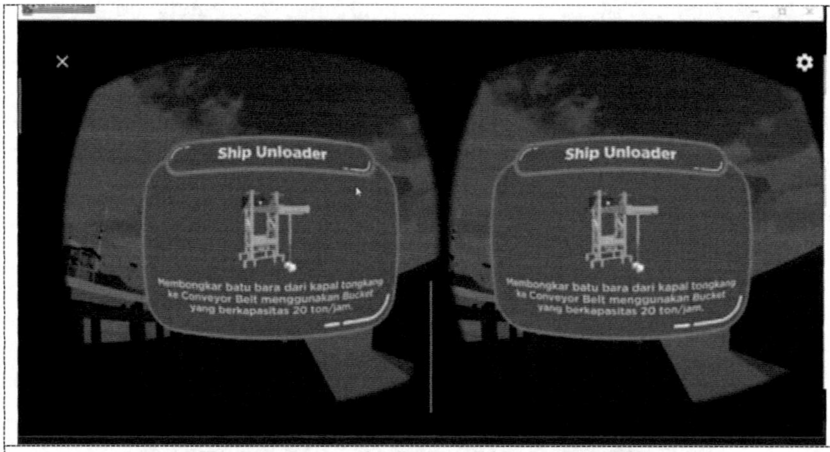

Figure 3. test the Power Plant application

Checking the execution of the Power Plant application using the questionnaire process, which is a checklist table that is entered explicitly by the user after attempting to execute the Power Plant application. The questionnaire you use applies to the USE Questionnaire. The questionnaire is prepared and given to 20 students in its sub-use, accessibility, quality of use, ease of learning and contentment. Several questions will be posed with the following answers: "Strongly agree," "Agree," "Disagree," "Strongly disagree." Any problems are designed such that whether a person is pleased with the process of education he or she expresses his or her approval. The example of a query is given in Table here.

IOP Conf. Series: Materials Science and Engineering 1071 (2021) 012019 doi:10.1088/1757-899X/1071/1/012019

Table 2. Questionnaire.

NO.	QUESTION	RESPONSE
1	I love studying electric power generation	3 strongly agree, 16 agree, 2 disagree
2	I am more excited about studying power plant with this app	2 strongly agree, 16 agree, 3 disagree
3	This application makes it easier for me to observe the power generation process	3 strongly agree, 15 agree, 3 disagree
4	How to use this application is very easy	5 strongly agree, 15 agree, 1 disagree
5	I like the look of this app	2 strongly agree, 15 agree, 4 disagree

The following table shows the results of the limited implementation of these 21 students at SMK Nurul Jadi Probolinggo. The implementation in the chart below starts from questions 1-5.

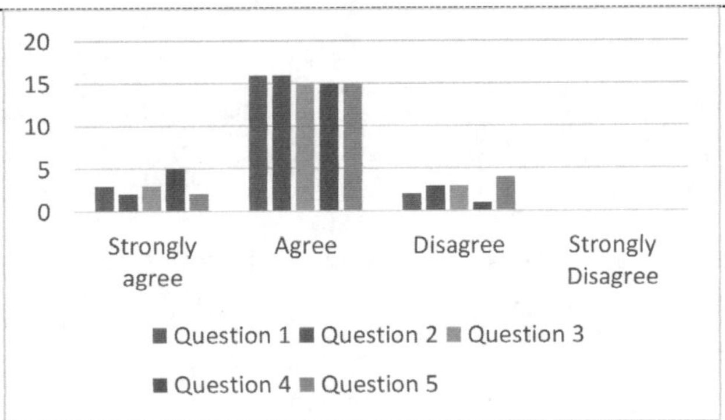

Figure 4. graph of the implementation results questionnaire

The limited questionnaire results show that the students agree with the use of the Power Plant VR application. It is hoped that it can be implemented in a wider environment in the future so that more data can be obtained for further better application development.

It is hoped that this technology will provide an easy way to achieve progress in the teaching and learning process of SMK Nurul Jadid Probolinggo students. So that it becomes an attraction as a learning medium in the form of animated 3D display media

4. Conclusion

Learning media that are boring and tend to be unattractive can make most people feel lazy and reluctant to know and explore a discipline such as what is needed by most students in Indonesia and even in Southeast Asia. Based on this, a learning media with virtual reality technology called Power Plant was created as a learning medium in one of the majors in the Vocational High School (SMK) in the power generation engineering department (TPTL). The design of this prototype is based on the Core Competencies (KI) and Basic Competencies (KD) of Vocational High Schools. However, there are many aspects that need to be addressed when using this technology, such as the creation of more attractive 3D models and simulations, gaming to draw student learning interest, and so on. This research also needs to be developed with respect to the model used not only for the learning of electrical energy, but it is hoped that it will also be developed for other topics.

References

[1] W. Ramansyah, N. Aini, I. Arkadiantika, and A. F. R. Nur Satiantoro, "Development of virtual reality technology in environmental pollution media learning content for junior high school students," *J. Phys. Conf. Ser.*, vol. 1511, no. 1, 2020, doi: 10.1088/1742-6596/1511/1/012026.

[2] L. Peng, Y. Yen, and I. Siswanto, "Virtual reality teaching material - Virtual reality game with education," *J. Phys. Conf. Ser.*, vol. 1456, no. 1, 2020, doi: 10.1088/1742-6596/1456/1/012039.

[3] A. K. Triatmaja and M. Khairudin, "Study on Skill Improvement of Digital Electronics Using Virtual Laboratory with Mobile Virtual Reality," *J. Phys. Conf. Ser.*, vol. 1140, no. 1, 2018, doi: 10.1088/1742-6596/1140/1/012021.

[4] J. Zhang, "Innovative application of virtual display technique in virtual museum," *IOP Conf. Ser. Mater. Sci. Eng.*, vol. 231, no. 1, 2017, doi: 10.1088/1757-899X/231/1/012043.

[5] I. Rizaldy, I. Agustina, and F. Fauziah, "Implementasi Virtual Reality Pada Tur Virtual Monumen Nasional Menggunakan Unity 3D Algoritma Greedy Berbasis Android," *JOINTECS (Journal Inf. Technol. Comput. Sci.*, vol. 3, no. 2, 2018, doi: 10.31328/jointecs.v3i2.786.

IOP Conf. Series: Materials Science and Engineering 1071 (2021) 012020 doi:10.1088/1757-899X/1071/1/012020

Web-based Design of Financial Apps: Case of Kosan 54

D Sari[1*], M Pradana[2], D W Nugraha[3] and F Oktafani[4]
[1,2,4]Telkom University, Indonesia
[3]Politeknik Pos Indonesia

E-mail: deviliasari@telkomuniversity.ac.id

Abstract. Financial Management is the process of financial planning in budgeting, auditing, management, control, deposit funds that are held for effective and efficient financial decisions.. As the existing financial management processes in boarding house number 54 (Kosan 54) is still manual, it is prone to mistakes during the recording and making their own bills and financial statements, along with the process of bill payment by the boarding members. Therefore, a system that can perform data processing in an integrated and well computerized system using a WEB-based system is urgently needed. The designed system is a WEB-based Financial Management Application. To build the application, a software such as MySQL database is needed, as well as using the PHP programming language. This finance management application then will be of great help in the planning of appropriate financial income and expenses incurred.

1. Introduction

Along with advances and developments in technology, the human need for technology becomes a very useful and common need. The use of technology is almost required in all aspects, including education, health, economy, and even in social, and culture. Currently, KOSAN 54, which consists of 10 members, is still using a bookkeeping system and manual calculations in terms of managing their finances, which includes their monthly income, expenses, and many others. Because of this, errors constantly occur in terms of calculations, bookkeeping, as well as their own financial reports.

Financial management is an activity that must be emphasized and considered by every organization, especially those which will help them achieve their goals or their desires in the future. For this reason, a website was created that could solve this problem. This website will make it easy for the board of Kosan 54, where they can access this site to manage their finances much easily so that they can be managed properly and without the need to use pens, calculators, paper, or other manual tools as they have been integrated and well computerized in the form of a website.

Based on the background of the problems that have been stated above, it can be identified that: one of the difficulties of the boarding house managers is knowing and informing the due date of bill payments to their boarding members. The financial recording is still carried out using a handwritten method so that it is less than optimal and effective

The purpose of making this application is to help the boarding house managers in reminding the payment due to boarding house members. Based on the above background, the discussion of this study is limited as follows: 1. Information from the web is only used to display billing information for Kosan 54 members. 2. The system only serves financial records by the boarding house managers.

IOP Conf. Series: Materials Science and Engineering 1071 (2021) 012020 doi:10.1088/1757-899X/1071/1/012020

2. Theoretical Basis

An information system is a system within an organization that meets the processing needs of daily transactions, supports operations, is managerial, with strategic activities of an organization and provides certain outsiders with the necessary reports [1].

While money is an object that can be exchanged for other objects, can be used to value other objects, and something we can save. Furthermore, money can also be used to pay debts in the future. In other words, money is an object that can basically function as: (1) a medium of exchange, (2) a store of value, (3) a unit of account, and (4) the measure of the deferred payment (standard for deferred payment) [2] All company activities related to the effort of obtaining company funds at low costs and the effort to use and allocate these funds efficiently [3].

A website or site can be interpreted as a collection of pages that display information on text data, still or motion image data, animation data, sound, video, or a combination of all of them. This includes both static and dynamic data which forms a series of interrelated buildings where each is linked by network pages (hyperlinks) [4] The definition of the internet is a series or network of a number of interconnected computers. The Internet comes from the word interconnected-networking. The Internet is a global network that connects a network with other networks around the world. The media that connects the network can be in a form of a cable, satellite channel, or radio frequency [5].

2.1. PHP (hypertext preprocessor)

It is According to the official PHP documents, PHP stands for Hypertext Preprocessor. It is a language in the form of scripts that are placed and processed on the server. The results will then be sent to the client, where the user uses the browser. [6] A database is a collection of information which is systematically stored in a computer. This allows it to be checked by a computer program to obtain information from the database. The software used to manage and call database queries is called a database management system (DBMS). [7] A flowmap is a picture in the form of a flow chart of an application program which states a direction or flow of the program [8] The analysis and the system design is the first step in system development. This is done to determine needs, problems that can be solved with the system that is being built, and what kind of system is going to be built [9].

3. Analysis

Analysis of the financial condition of boarding house 54 which starts from the payment process to the recording process of observations and analyzes that have been carried out, the current business flow or flowmap in financial management at Kosan 54 is as follows:

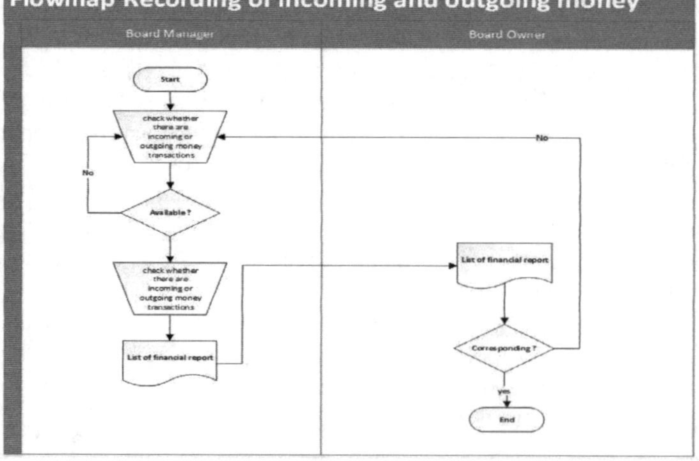

Figure 1. Current Flowmap of Recording Money In and Out

3.1. Design

A context diagram is a simple diagram that describes the relationship between the external entities, input, and output from the system. A context diagram is represented by a single circle that represents the whole system.

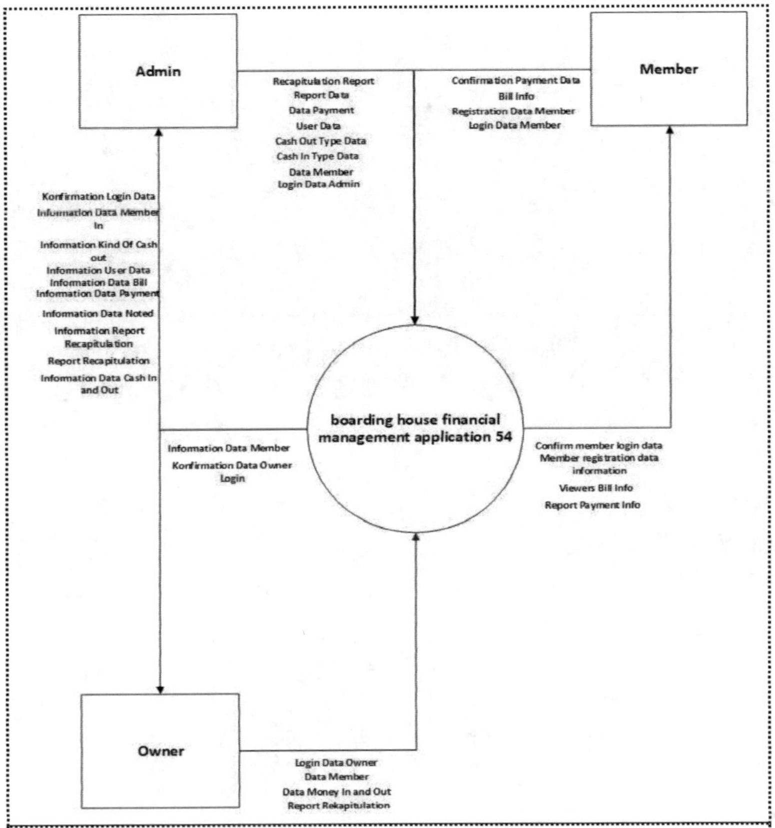

Figure 2. Context Diagram

4. Implementation Environment

After analyzing and designing the system that is going to be built,, the system is then ready to be implemented.

This stage is where the system is being placed so that it can be operated in accordance with the analysis that has been made. The programming and adjustments to the designs that have been done before are also carried out in this stage.

4.1. Administrator Login Page

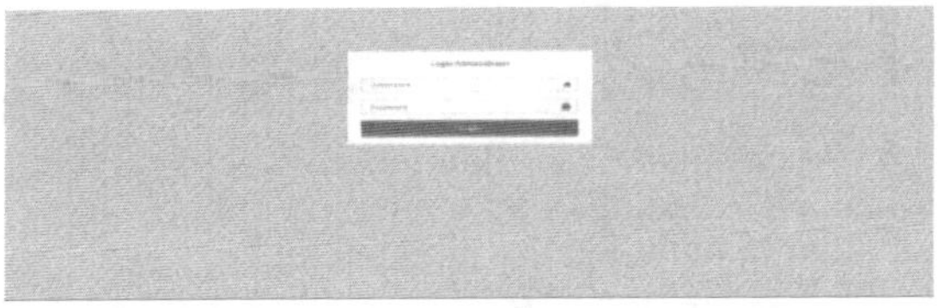

Figure 3. Administrator Login Page

Instructions for use:
1. Displays a page that contains a form so that managers can enter the main page and operate the system.
2. Click the desired menu to open the page.

4.2. Main Admin Page

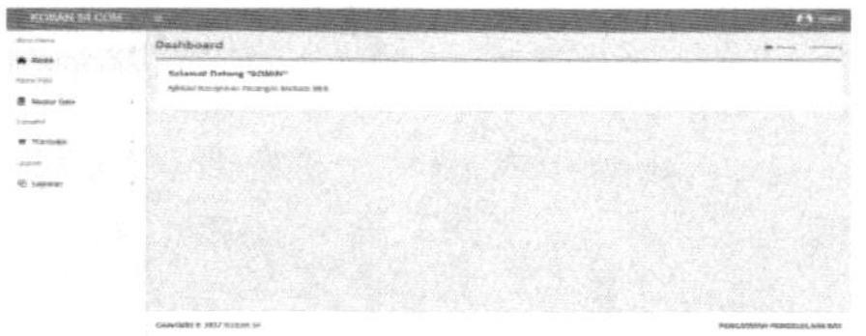

Figure 4. Main Admin Page

Instructions for use:
1. Enter the website address.
2. Click the desired menu to open the page.

4.3. Member Login Menu

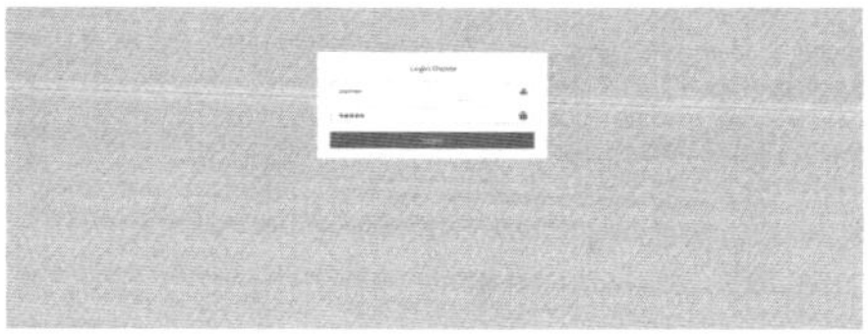

Figure 5. Member Login Menu

Instructions of or use:
1. Enter the website address.
2. Click the desired menu to open the page.

4.4. Registration Page

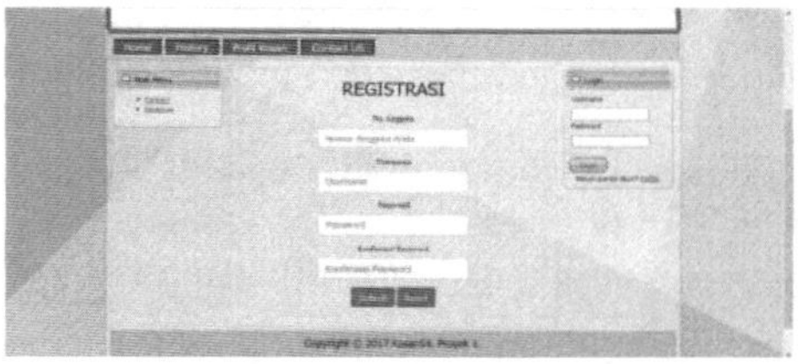

Figure 6. Registration Page

IOP Conf. Series: Materials Science and Engineering 1071 (2021) 012020 doi:10.1088/1757-899X/1071/1/012020

Instructions for use:
1. Displays a page that contains a form for registration to log into the system.
2. Click main registration menu to open registration or the link "register" under login menu.

4.5. Member Home Page

Figure 8. Billing Info Page

Instructions for use:
1. Displays a page that contains bills that must be paid by the boarding house members after logging in.
2. Click the billing info menu bar to open the page.

4.6. Payment Page

Figure 9. Payment Page

Instructions for use:
1. Displays a page that contains a form to confirm payment;
2. Click the 'pay' menu button.
3. Fill in the payment details.
4. Click the Submit button to confirm.

5. Conclusion

After analyzing and designing the system that is going to be built,, the system is then ready to be implemented. This stage is where the system is being placed so that it can be operated in accordance with the analysis that has been made. The programming and adjustments to the designs that have been done before are also carried out in this stage The application built has made it easier to confirm the payment of bills by boarding house members of Kosan 54. The application that was built will also facilitate the control process in managing computerized information and finance. From the conclusions above, there are some suggestions that are expected to help and overcome the shortcomings of this application. The suggestions is that the update of this application should develop new features that can make the application easier to use, namely: Adding new features such as a more complete payment system which includes an annual booking option and booking payments. Also payment for electricity, water, Wi-Fi, and the process of extending or choosing to stop renting the boarding rooms. This application is not yet able to display the data of which rooms that have not been ordered. It also does not include the types of rooms. These shortcomings can be considered in the development process of the application to the next stage.

References

[1] Pradana, M., & Novitasari, F. (2017). Gap analysis of Zalora online application: Indonesian users' perspectives. International Journal of Learning and Change, 9(4), 334-347.
[2] Saraswati, T. G., & Basri, M. H. (2016). Simulation model for evaluating intensive care unit capacity. Актуальні проблеми економіки, (3), 414-420.
[3] Saragih, R., Fakhri, M., Pradana, M., Gilang, A., & Vidjashesa, G. A. (2018). Ethical Leadership's Effect on Employee Discipline: Case of An Indonesian Telecommunication Company. In Proceedings of the International Conference on Industrial Engineering and Operations Management.
[4] Silvianita, A., Tan, C.-L. (2017). A model linking the knowledge management (KM) enabler, KM capability and operational performance in indonesian automobile industry. Advanced Science Letters, 23(1), pp. 640-642
[5] Pradana, M., Wahyuddin, S., Syarifuddin, S., & Putra, A. Gap Analysis of Indonesian State-Owned Bank Internet Banking Website.
[6] Hidayat, R., & Hidayat, A. M. (2017). Contribution of Environment Online Store for Purchase Decision (A Case Study Online Store Hypermart—Bandung). Advanced Science Letters, 23(1), 292-294.
[7] Hafifah, D. K., Witarsyah, D., Saputra, M., Azizah, A. H., & Saputri, M. E. (2019, September). The Evaluation of Finance Module Implementation of Enterprise Resource Planning (ERP) for Employee Performance. In 2019 2nd International Conference on Applied Information Technology and Innovation (ICAITI) (pp. 140-145). IEEE.
[8] Pradana, M., & Wijaksana, T. I. (2017). Managing work productivity through management of information system (Study on Telkom Indonesia Online Portal). Advanced Science Letters, 23(1), 64-66.
[9] Priharti, W., Sumaryo, S., Saraswati, T., & Nurfadilah, M. R. (2020). IoT Based Logistics Vehicle Security Monitoring System. MS&E, 771(1), 012012.

IOP Conf. Series: Materials Science and Engineering 1071 (2021) 012021 doi:10.1088/1757-899X/1071/1/012021

Neural Collaborative For Music Recommendation System

Abba Suganda Girsang[1*], Antoni Wibowo[1], Jason[1] and Roslynlia[1]
[1]Computer Science Department, BINUS Graduate Program – Master of Computer Science, Bina Nusantara University, Jakarta, 11480, Indonesia

E-mail: agirsang@binus.edu

Abstract. In music there are various types of genres and every person has their own choice of the type of music they want to hear. Recommendation system is an important feature in an application, especially with the large number of choices for a particular item. With a good recommendation system, users will be helped by the suggestions given and can improve the user experience of the application. It is better provided by using collaborative filtering (CF) approach by recommending products related to one's preferences history. However, CF approach still lacking in integrating complex users data. Therefore, hybrid technique could be the solution to polish the CF approach. Combining neural network and CF also called NCF thought to be better than CF alone. This study focuses on collaborative filtering approach combined with neural network or called neural collaborative filtering. In this study, we use 20,000 users, 6,000 songs, and 470,000 transaction ratings then predict the score using CF and NCF approach. The aim of this study is to differentiate recommendation systems with the use of CF alone and NCF. Through this research, it was found that NCF is better than user-based collaborative filtering in gather those playlist they really want to hear, but requires more time to build it.

1. Introduction

Information related to entertainment is one of the most sought after things in various digital information media. Content that shares information in the form of music is very easy to obtain. Referring to these things, of course, digital content has become one of the important needs to be provided in the form of attractive information with various methods. Generally, digital music content providers will take advantage of user behavior data when exploring the digital world via the internet. Currently, millions of song catalogs have been distributed on the internet and can be accessed easily by users.

On average, each digital-based song content has a duration of around 3 minutes, so it can be concluded that with that duration to complete 1 million digital music content it will take approximately 5 years. The increase in the amount of digital music content every month results in an unstructured amount of song catalog data and makes it difficult for users to choose the songs they want to listen to. To make it easier for users to optimize the music catalog that has been subscribed to a large number, a music recommendation-based system is needed that allows users to be able to manage a catalog of digital music content according to their needs. One of the method recommendation is collaborative filtering. Collaborative screening recommendation techniques began to emerge in the decade of the 90's mainly due to the increasing number of web and the vast amount of information available through it. These techniques are domain independent and use information from multiple users to determine similarities between items. In collaborative filtering, the most popular

IOP Conf. Series: Materials Science and Engineering 1071 (2021) 012021 doi:10.1088/1757-899X/1071/1/012021

technique is matrix factorization (MF), but it is still lacking in capturing the complex structure of user interaction data. Therefore, the need of combining another technique is required to give a better outcome [1]. Currently, hybrid techniques are the ones most extensively implemented in recommender systems in an attempt to address the limitation of CF approach [2], [3]. One of which is by combining with neural networks for learning the interaction function from data [1]. This study focuses on collaborative filtering approach combined with neural network or called neural collaborative filtering (NCF). This study aims to enhance the collaborative

2. Related Work

Collaborative recommendation system relied on two different types of input, of which are explicit feedback and implicit feedback. In the explicit feedback, data obtained from input by users regarding their interest in products. However, it is not always available. Therefore, we need the implicit feedback by observing user behavior. Some examples of implicit feedback are mouse click, number of times a video or music played, etc. Note to the implicit feedback, it has no negative feedback, it is inherently noisy, the numerical value indicates confidence, and its evaluation requires appropriate measures [4].

The basic needs to create a neighbourhood method in CF recommendation system consist of identifying similar users based on their preferences, then select k nearest neighbour based on users similarity [1], [5], [6]. The top k nearest neighbour then gives a good quality of recommendation. The other and most popular method is matrix factorization (MF) that reflects users and items using latent vector [7]. Another research showed that processing the data with a neighbourhood method using KNN algorithm gives better error result as well as good to predict similarities between the pair of song given the attributes [2], [6].

However, each technique had its own weakness, therefore combining techniques could give a promising future in providing better service. Currently, hybrid techniques are the ones most extensively implemented in recommender systems in an attempt to address the limitation of CF approach [2], [3].

3. Methodology

Neural collaborative filtering approaches combine general matrix factorization with neural network matrix factorization as shown Figure 1. Each transactions are extracted into two matrices / part named user vector, and item vector. The user vector and item vector are then represented into generalized matrix factorization (GMF). In other hand, two vectors user and item are concatenated into multi layer perceptron (MLP). , then we create four dense layers and each layer using dropout with rate 0.25. Each dense layer is using ReLu activation function. Next, we concatenate general matrix factorization with the output of the neural network layer. Finally, the output layer is using ReLu activation function as we want to predict the music rating. The formulation of neural collaborative filtering as shown in Eq.(1), Eq(2), Eq(3) [1]

$$GMF = p_u \cdot q_i \tag{1}$$

$$\mathrm{MLP} = \alpha_L\left(W_L^T\left(\alpha_{L-1}\left(\ldots\alpha_2\left(W_L^T\begin{bmatrix}p_u\\ q_i\end{bmatrix}+b_2\right)..\right)\right)+b_L\right) \tag{2}$$

$$y_{ui} = \sigma\left(h^T\begin{bmatrix}GMF\\ MLP\end{bmatrix}\right) \tag{3}$$

Where p_u denotes user embedding and q_i denotes item embedding. The training model is linear regression and the loss function is mean square error (MSE). Parameters are batch size = 64, epochs = 10, learning rate = 0.001, and use ADAM optimizer.

Data used in this study consist of users, music, and the total number of times the music was played. There are over 20,000 users, 6,105 music, and 470,759 records of music played. We need to normalize the number of times the music is played c_{ui} due to different habits each user is likely listening to music. The normalize value then considered as rating r_{ui} – the rating from user u for item I as shown Eq. (4):

$$r_{ui} = \frac{c_{ui}}{Max(c_u)} \times 100 \quad (4)$$

Rating value is then stored in the database and used for suggesting music content to each user. It is ranging from 0-100.

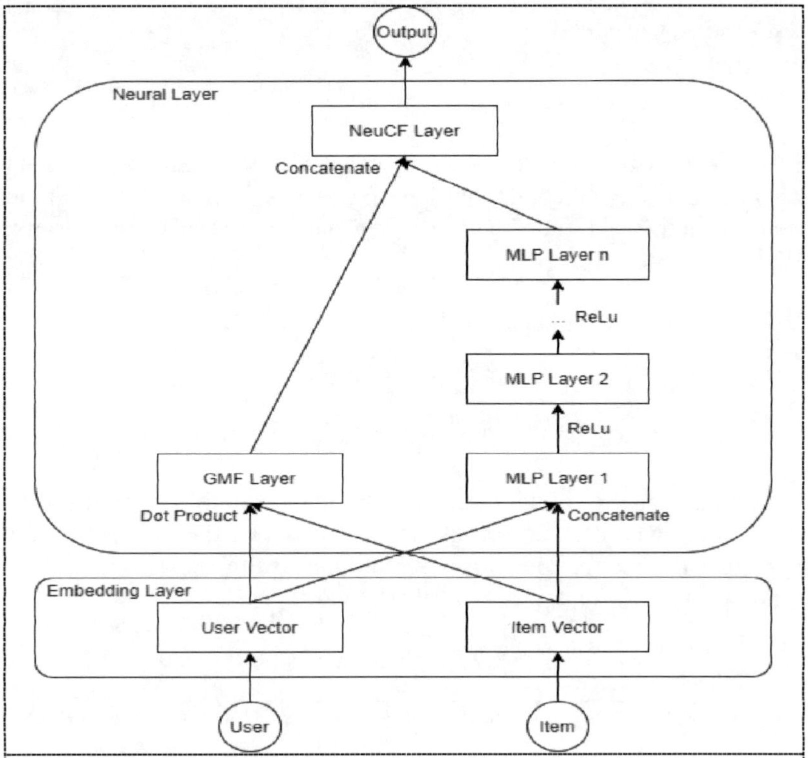

Figure 1. Architecture of Neural Collaborative Filtering.

4. Evaluation

To evaluate and compare our music recommendation systems, we used regression evaluation metrics: mean absolute error (MAE), mean absolute percentage error (MAPE), mean square error (MSE), and root mean square error (RMSE). Regression evaluation metrics are used due to the output of the recommendation system is the predicted rating. Thus, it calculates the difference between actual and predicted value of music rating. The formula of each evaluation metrics as shown Eq. (5)- Eq(8)

$$\text{MAE}=\frac{\sum_{i=1}^{n}|y_i - x_i|}{n} \quad (5)$$

$$\text{MAPE}=\frac{\sum_{i=1}^{n}\frac{|y_i - x_i|}{y_i}}{n} \times 100\% \quad (6)$$

$$\text{MSE}=\frac{\sum_{i=1}^{n}(y_i - x_i)^2}{n} \quad (7)$$

$$\text{RMSE}=\sqrt{\frac{\sum_{i=1}^{n}(y_i - x_i)^2}{n}} \quad (8)$$

Where y and x denote actual and predicted value respectively.

Based on the methods and steps described in chapter 3, the results of evaluation metrics of collaborative filtering and neural collaborative filtering can be seen in Table 1. From that results, it can be seen that the error from the NCF framework is lower than user-based CF using several methods like MAE, MAPE, MSE, and RMSE.

Table 1. Evaluation Error Results

Method	MAE	MAPE	MSE	RMSE
CF	14.866	66.118	645.786	26.107
NCF	7.933	28.591	246.422	15.697

Table 2 shows several examples of the results of recommendations using collaborative filtering and neural collaborative filtering. It uses 2 users given 7 songs, then the CF recommendation and NCF recommendation predict the rating. The closest prediction to the actual rating is the best. A better predictive score is done by NCF framework with 11/14 score.

Table 2. Evaluation Results

# User	# Song	Actual	CF Score	NCF Score	Winner
1	1	34	43.666	25.124	NCF
	2	13	28.5	10.368	NCF
	3	20	54.75	40.449	NCF
	4	17	13.3	18.745	CF
	5	17	19.66	22.226	CF
	6	3	27	25.456	NCF
	7	6	26.8	25.338	NCF
2	1	5	20.4	17.052	NCF
	2	50	58	51.9	NCF
	3	20	54.75	40.449	NCF
	4	5	13.5	13.961	CF
	5	5	17.666	6.296	NCF
	6	5	23.375	6.747	NCF

7	11	85	34.424	NCF

Based on Table 1 and 2, it showed that NCF framework generates less error and more accurate predicted rating. Unfortunately, matrix factorization's model using NCF needs lots of time to build, yet it predicts ratings more faster if we use the model that was built before than user-based CF.

In the application of music recommendation, the system will perform user-based collaborative filtering calculations, it is different from neural collaborative filtering which only predicts the model that has been built. This gives an advantage in run time because it does not require a long training process, but in real application it must be monitored and provided feedback so that the model still gives the best performance. The comparison of the two frameworks can be seen in Table 1, and Table 2.

5. Conclusion

In this study, we make a comparison between collaborative filtering and neural collaborative filtering on the music dataset. From the comparison, it was found that neural collaborative filtering gave better results than collaborative filtering. This process can increase people's engagement with an application, especially digital online music, because the right recommendations can improve the user experience of an application. The next step of this study is to improve recommendation performance so that existing errors can be reduced and produce better recommendations.

6. References

[1] X. He, L. Liao, H. Zhang, L. Nie, X. Hu, and T. S. Chua, "Neural collaborative filtering," arXiv, 2017.

[2] D. Jayashree, S. Goutham Manian, and C. Pranav Srivatsav, "Music recommendation system," Asian J. Inf. Technol., vol. 15, no. 21, pp. 4250–4254, 2016.

[3] D. Sánchez-Moreno, A. B. Gil González, M. D. Muñoz Vicente, V. F. López Batista, and M. N. Moreno García, "A collaborative filtering method for music recommendation using playing coefficients for artists and users," Expert Syst. Appl., vol. 66, pp. 1339–1351, 2016.

[4] Y. Hu, C. Volinsky, and Y. Koren, "Collaborative filtering for implicit feedback datasets," Proc. - IEEE Int. Conf. Data Mining, ICDM, pp. 263–272, 2008.

[5] A. K. Azmi, N. Abdullah, and N. A. Emran, "A hybrid knowledge-based and collaborative filtering recommender system model for recommending interventions to improve elderly wellbeing," Int. J. Adv. Trends Comput. Sci. Eng., vol. 9, no. 4, pp. 4683–4689, 2020.

[6] S. Ayyaz and U. Qamar, "Improving collaborative filtering by selecting an effective user neighborhood for recommender systems," Proc. IEEE Int. Conf. Ind. Technol., vol. 1, no. 2, pp. 1244–1249, 2017.

[7] N. Sivaramakrishnan, V. Subramaniyaswamy, S. Arunkumar, A. Renugadevi, and K. K. Ashikamai, "Neighborhood-based approach of collaborative filtering techniques for book recommendation system," Int. J. Pure Appl. Math., vol. 119, no. 12, pp. 13241–13250, 2018.

Utilization of the Batch Training Method for Predicting Natural Disasters and Their Impacts

N L W S R Ginantra [1], M A Hanafiah [2], A Wanto [3*], R Winanjaya[3] and H Okprana[3]
[1] STMIK STIKOM Indonesia, Denpasar, Bali, Indonesia
[2] Politeknik Bisnis Indonesia, Pematangsiantar, Indonesia
[3] STIKOM Tunas Bangsa, Medan, Indonesia

E-mail: anjarwanto@amiktunasbangsa.ac.id

Abstract. Indonesia is one of the countries that often experiences natural disasters, including earthquakes, floods, tsunamis, etc. All of this causes losses, both casualties, Broken, and Anguishing for the population. Based on this, this paper is proposed, which aims to predict natural disasters in the coming years in Indonesia, casualties, Broken, and their consequences. This paper is an extension of previous research, which is still an architectural model to predict Indonesia's natural disasters and their impacts. Model 4-10-1 is the best in this study, which produces 91% accuracy. Based on this architectural model, this paper will predict natural disasters that occur and their impacts for the years to come in Indonesia. The research dataset and algorithms used remain the same, namely the natural disaster dataset for 2008-2019. Resourced from its National Emergency Management Department and the Batch Training algorithm. Specifically, the results of this proposed paper are in the form of a prediction of natural disasters that will occur, dead and disappear, injured, Anguishing and displaced, houses severely Broken, moderately Broken, lightly Broken to submerged, and Broken to facilities and infrastructure such as health facilities, facilities. worship and educational facilities.

1. Introduction

The occurrence of natural disasters is usually triggered by a condition or series of natural events such as floods, tsunamis, earthquakes, volcanic eruptions, landslides, hurricanes, drought and others. In general, natural disasters arise due to changes in nature, both gradual and extreme. But there are also natural disasters that arise because of human activity, such as logging that can cause landslides. Many researchers and scientists to date have tried various ways so that natural disasters can be detected. However, in reality, natural disasters are always a mystery because identification and detection are quite difficult. Almost every natural disaster that occurs is bound to cause loss, damage to casualties. In addition, natural disasters can influence subjective perceptions about the seriousness of potential disasters and their prevalence [1]. The impact of natural disasters also has implications for people's lives, especially in the social sector [2]. Experts and analysts also argue that the occurrence of natural disasters has a very negative effect on state policy [3]. Some of the world's natural disasters have caused significant harm to the social building and economic growth, including the 2004 Indonesian tsunami and the 2008 Wenchuan earthquake [4], Southern China's freezing rain disaster in 2008, Japan's earthquake disaster in 2011, India's flood disaster in 2013 and Jiangsu hail in Yancheng in

2016. Continues to hit Heavy rains in July 2016 in North China, along with major floods that hit South China [5].

According to the National Emergency Management Department of Republic Indonesia, natural disasters in 2008-2018 in the last decade have increased quite a bit, but in 2019 this number decreased as recorded from 3,406 to 1,999 alone. Based on statistics from the National Emergency Management Department, natural disasters in 2018 were the biggest disasters compared to previous years, due to the large number of natural disasters, depression and loss of homes, dead and missing victims, and broken moderate houses [6].

Table 1. Natural Disasters Data in Indonesia

Times	Total	Toll (soul)			House (unit)				Broken (unit)		
		Dead & Disappear	Injuries	Anguish & Evacuate	Broken Severely	Broken Moderate	Broken Lightly	Submerged	Facility Medical	Facility Worship	Facility Educational
2010	1.947	1.907	35.730	1.663.103	20.084	3.709	35.708	686.523	367	628	1.557
2011	1.622	428	692	475.529	13.549	3.358	56.736	194.785	106	457	566
2012	1.781	320	1.198	954.241	10.686	8.018	35.356	187.601	89	219	473
2013	1.666	512	3.410	3.892.986	17.727	12.590	59.401	470.756	321	684	1.902
2014	1.963	604	2.104	2.814.265	20.079	6.067	29.350	525.434	63	355	766
2015	1.694	276	370	1.227.929	5.217	3.871	16.444	180.319	33	159	309
2016	2.308	578	2.675	3.162.491	9.029	9.979	28.790	334.606	232	601	1.484
2017	2.868	378	1.042	3.674.369	10.452	10.648	28.631	376.373	117	715	1.326
2018	3.406	2.046	19.610	10.364.764	117.655	70.303	182.195	313.653	287	1.176	2.984
2019	1.999	367	1.431	651.095	4.047	4.181	16.857	153.025	102	281	452

Because disaster information is so important as an Early Warning step for the people of Indonesia, it is necessary to predict the number of disasters in the following year as well as an estimate of casualties (Death and Missing, Injured, Anguished and Displaced), Houses / Units (Severely Broken, Moderately Broken) , slightly Broken and submerged) and Broken / unit (health facilities, worship facilities and educational facilities). One of the good algorithms used to make predictions is the batch training algorithm. This algorithm is one of the Algorithms of the ANN. Artificial Neural Networks are widely used for solving problems related to estimation (prediction), pattern recognition, data analysis, control and grouping [7]–[12].

This research was conducted as a development of previous research conducted by Ginantra, et al (2020). As it should be noted that the results of this study are still in the form of a network architecture model to predict natural disasters and their impact using Batch training, not yet at the stage of the resulting predictive value [13]. Therefore in this paper, a calculation process will be carried out to obtain a prediction result in the form of the number of disasters, casualties (Death and Disappear, Injuries, Anguishing, and Evacuate), Houses/Units (Broken Severely, Broken Moderate, Broken lightly and submerged) or Broken/unit (Facility Medical, Facility Worship, and Facility Educational). The next research was conducted by Wanto, et al. (2019) to analyze the Batch Training method's accuracy in seeing the development of Indonesian aquaculture companies. The result is that the accuracy of this method reaches 75%, which means it is good enough to use [14].

Based on this background, a study was conducted to predict natural disasters and their impacts using a batch training algorithm. This study's results are expected to be input and reference for the government to determine policies or make appropriate strategic steps to respond to disasters that will occur in Indonesia for the coming years to minimize casualties and Broken.

IOP Conf. Series: Materials Science and Engineering 1071 (2021) 012022 doi:10.1088/1757-899X/1071/1/012022

2. Methodology

2.1. Techniques and Methods
Collecting data uses quantitative techniques. While the method used is batch training (trainb). This approach will render data-based predictions (times series). The weighting and bias processes in accordance with the learning function with this method will be modified.

2.2. Techniques and Methods
The stages carried out in this study can be seen in Figure 1.

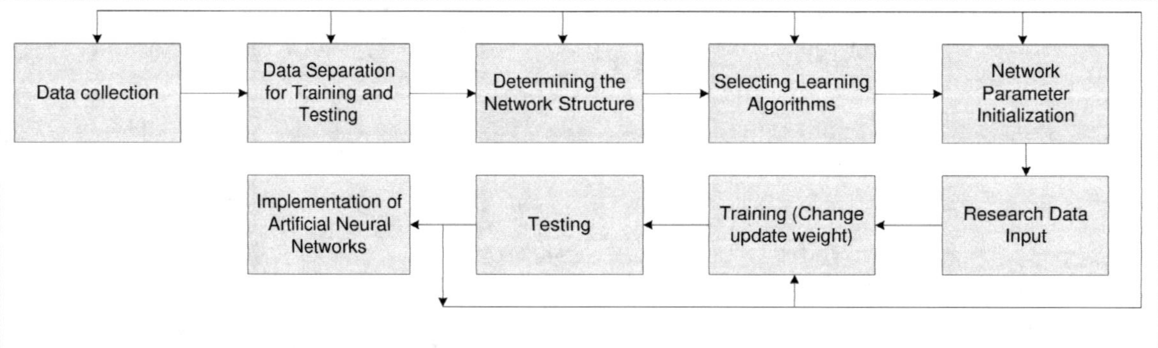

Figure 1. Research Stages

2.3. Variables
The twelve variables of this analysis are the time of catastrophe, the total of disasters, dead and Disappear, Injuries, Anguishing and Evacuate, seriously Broken houses, lightly Broken houses, Submerged, Broken facility medical, Broken facility worship, and Broken facility educational.

3. Results and Discussion

3.1. Normalization
Based on the data in Table 1, first normalization will be carried out using equations (1) berikut: [15]–[18].

$$x' = \frac{0.8(x-a)}{b-a} + 0.1 \tag{1}$$

Table 2. Normalization Results

Years	Total	Toll (Soul)			House (unit)					Broken (unit)		
		Dead & Disappear	Injuries	Anguish & Evacuate	Broken Severely	Broken Moderate	Broken Lightly	Submerged		Facility Medical	Facility Worship	Facility Educational
2010	0.10015	0.10014	0.10276	0.22836	0.10155	0.10028	0.10275	0.15299		0.10003	0.10005	0.10012
2011	0.10012	0.10003	0.10005	0.13670	0.10104	0.10026	0.10438	0.11503		0.10001	0.10003	0.10004
2012	0.10013	0.10002	0.10009	0.17365	0.10082	0.10062	0.10273	0.11448		0.10000	0.10001	0.10003
2013	0.10013	0.10004	0.10026	0.40048	0.10137	0.10097	0.10458	0.13633		0.10002	0.10005	0.10014
2014	0.10015	0.10004	0.10016	0.31722	0.10155	0.10047	0.10226	0.14055		0.10000	0.10002	0.10006
2015	0.10013	0.10002	0.10003	0.19477	0.10040	0.10030	0.10127	0.11392		0.10000	0.10001	0.10002
2016	0.10018	0.10004	0.10020	0.34409	0.10069	0.10077	0.10222	0.12582		0.10002	0.10004	0.10011
2017	0.10022	0.10003	0.10008	0.38360	0.10080	0.10082	0.10221	0.12905		0.10001	0.10005	0.10010

		Toll (Soul)				House (unit)			Broken (unit)		
Years	Total	Dead & Disappear	Injuries	Anguish & Evacuate	Broken Severely	Broken Moderate	Broken Lightly	Submerged	Facility Medical	Facility Worship	Facility Educational
2018	0.10026	0.10016	0.10151	0.90000	0.10908	0.10542	0.11406	0.12421	0.10002	0.10009	0.10023
2019	0.10015	0.10003	0.10011	0.15025	0.10031	0.10032	0.10130	0.11181	0.10001	0.10002	0.10003

3.2. Results of Training and Testing

Based on previous research [13], There are 3 architectural models used: 4-5-1, 4-10-1, and 4-15-1. Model 4-10-1 is the best because of its 91% accuracy rate (higher than other architectural models). The parameters used are also the same as the study. The analysis process uses Matlab and Microsoft Excel tools. The training and testing results of the three models used can be seen in Table 3, Table 4, Table 5, Table 6, Table 7, and Table 8.

Table 3. Training with Model 4-5-1

P	T	O	E	SSE
P1	0.10015	0.12560	0.02545	0.0006477551
P2	0.10004	0.12620	0.02616	0.0006841325
P3	0.10016	0.12830	0.02814	0.0007918681
P4	0.31722	0.73840	0.42118	0.1773959463
P5	0.10155	0.12150	0.01995	0.0003981123
P6	0.10047	0.12410	0.02363	0.0005585786
P7	0.10226	0.10720	0.00494	0.0000243757
P8	0.14055	0.30540	0.16485	0.0271745377
P9	0.10000	0.12620	0.02620	0.0006863187
P10	0.10002	0.12610	0.02608	0.0006799133
P11	0.10006	0.12610	0.02604	0.0006782599
				0.2097197982
			MSE	0.0190654362

Table 4. Testing with Model 4-5-1

P	T	O	E	P	Results
P1	0.10015	0.12520	0.02505	0.0006274151	0
P2	0.10003	0.12600	0.02597	0.0006746601	0
P3	0.10011	0.12480	0.02469	0.0006096996	0
P4	0.15025	0.90700	0.75675	0.5726673694	0
P5	0.10031	0.11960	0.01929	0.0003721110	1
P6	0.10032	0.12020	0.01988	0.0003952079	1
P7	0.10130	0.11160	0.01030	0.0001061197	1
P8	0.11181	0.47510	0.36329	0.1319805966	0
P9	0.10001	0.12610	0.02609	0.0006809320	0
P10	0.10002	0.12600	0.02598	0.0006750050	0
P11	0.10003	0.12560	0.02557	0.0006537052	0
				0.7094428217	27%
			MSE	0.0644948020	

Table 5. Training with Model 4-10-1

P	T	O	E	SSE
P1	0.10015	0.10170	0.00155	0.0000024057
P2	0.10004	0.10110	0.00106	0.0000011150
P3	0.10016	0.09740	-0.00276	0.0000076168
P4	0.31722	0.83600	0.51878	0.2691368195
P5	0.10155	0.10530	0.00375	0.0000140832
P6	0.10047	0.10280	0.00233	0.0000054488
P7	0.10226	0.11600	0.01374	0.0001887099
P8	0.14055	0.16750	0.02695	0.0007261415
P9	0.10000	0.10110	0.00110	0.0000012049
P10	0.10002	0.10120	0.00118	0.0000013810
P11	0.10006	0.10120	0.00114	0.0000013074
				0.2700862337
			MSE	**0.0245532940**

Table 6. Testing with Model 4-10-1

P	T	O	E	SSE	Results
P1	0.10015	0.10220	0.00205	0.0000041953	1
P2	0.10003	0.10140	0.00137	0.0000018885	1
P3	0.10011	0.10240	0.00229	0.0000052537	1
P4	0.15025	0.94830	0.79805	0.6368804351	0
P5	0.10031	0.10550	0.00519	0.0000269380	1
P6	0.10032	0.10610	0.00578	0.0000334065	1
P7	0.10130	0.11460	0.01330	0.0001769284	1
P8	0.11181	0.10430	-0.00751	0.0000563800	1
P9	0.10001	0.10130	0.00129	0.0000016762	1
P10	0.10002	0.10140	0.00138	0.0000019068	1
P11	0.10003	0.10190	0.00187	0.0000034882	1
				0.6371924966	91%
			MSE	**0.0579265906**	

Table 7. Training with Model 4-15-1

P	T	O	E	SSE
P1	0.10015	0.10540	0.00525	0.0000275734
P2	0.10004	0.10470	0.00466	0.0000216777
P3	0.10016	0.10300	0.00284	0.0000080665
P4	0.31722	0.18450	-0.13272	0.0176135416

Table 8. Testing with Model 4-15-1

P	T	O	E	SSE	Results
P1	0.10015	0.10590	0.00575	0.0000330424	1
P2	0.10003	0.10480	0.00477	0.0000227932	1
P3	0.10011	0.10520	0.00509	0.0000259294	1
P4	0.15025	0.91380	0.76355	0.5830053807	0

Table 7. Training with Model 4-15-1

P	T	O	E	SSE
P5	0.10155	0.10990	0.00835	0.0000697685
P6	0.10047	0.10710	0.00663	0.0000440135
P7	0.10226	0.13160	0.02934	0.0008606697
P8	0.14055	0.11230	-0.02825	0.0007982313
P9	0.10000	0.10470	0.00470	0.0000220682
P10	0.10002	0.10470	0.00468	0.0000218570
P11	0.10006	0.10470	0.00464	0.0000215614
				0.0195090286
			MSE	0.0017735481

Table 8. Testing with Model 4-15-1

P	T	O	E	SSE	Results
P5	0.10031	0.10680	0.00649	0.0000421224	1
P6	0.10032	0.10850	0.00818	0.0000669097	1
P7	0.10130	0.11600	0.01470	0.0002161324	1
P8	0.11181	0.22650	0.11469	0.0131541031	0
P9	0.10001	0.10480	0.00479	0.0000229889	1
P10	0.10002	0.10490	0.00488	0.0000238228	1
P11	0.10003	0.10530	0.00527	0.0000277482	1
				0.5966409733	82%
			MSE	0.0542400885	

Explanation: P = Pattern, T = Target, O = Output, E = Error

Based on the data in Table 3, Table 4, Table 5, Table 6, Table 7, and Table 8, the results show that Model 4-10-1 is the best model chosen.

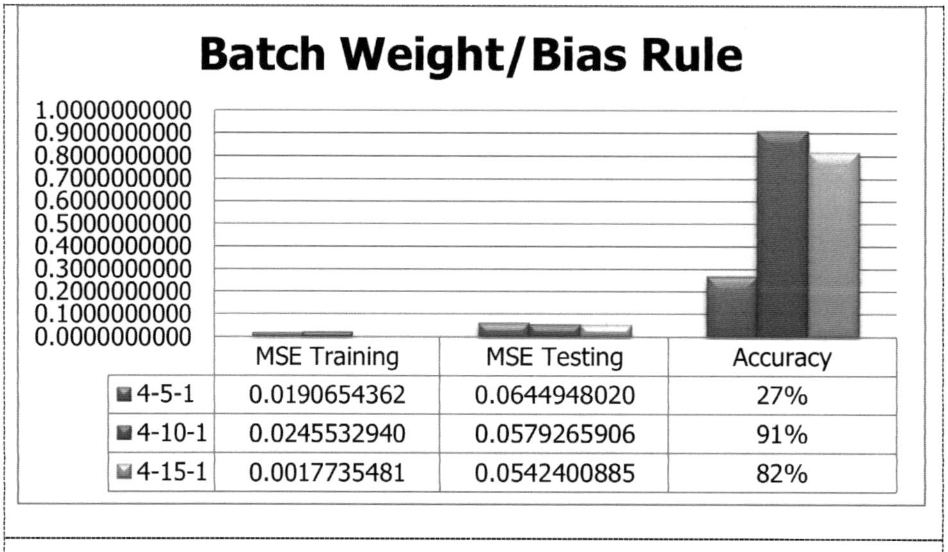

Figure 2. Comparison Graph of Accuracy Level and MSE Value

3.3. Prediction results

Then the prediction will be made using the 4-10-1 model using the formula to return the value:

$$x_n = \frac{(x - 0,1) * (b - a)}{0,8} + a \quad (2$$

Explanation:
x_n = Prediction Results
x = Predicted Target
a = The smallest data from the dataset
b = The largest data set from the dataset
For the prediction results in 2020 can be seen in Table 9.

Table 9. Results of Prediction of Natural Disasters in Indonesia (2020)

Years	Total	Toll (soul)			House (unit)				Broken (unit)		
		Dead & Lost	Injuries	Anguish & Evacuate	Severely Broken	Moderately Broken	Broken Lightly	Submerged	Medical Facility	Worship Facility	Educational Facilities
2016	2.308	578	2.675	3.162.491	9.029	9.979	28.790	334.606	232	601	1.484
2017	2.868	378	1.042	3.674.369	10.452	10.648	28.631	376.373	117	715	1.326
2018	3.406	2.046	19.610	10.364.764	117.655	70.303	182.195	313.653	287	1.176	2.984
2019	1.999	367	1.431	651.095	4.047	4.181	16.857	153.025	102	281	452
2020	**223**	**138**	**232**	**176.860**	**1.361**	**481**	**1.943**	**33.102**	**104**	**122**	**148**

4. Conclusion

The 4-10-1 architectural model with the Batch training method can be used to predict natural disasters in Indonesia with an accuracy rate of 91%. The architectural model and parameters used have an effect on the resulting level of accuracy. Based on the comparison between the Preliminary Data and the Prediction data (2020), the number of disasters and their consequences decreased quite significantly. However, in reality, until the end of 2020, natural disasters and their consequences has indeed decreased, but not significantly compared to the predicted data. Thus the results of this study need to be reviewed to obtain accurate results (even better).

References

[1] P. Brown, A. J. Daigneault, E. Tjernström, and W. Zou, "Natural disasters, social protection, and risk perceptions," *World Development*, vol. 104, pp. 310–325, 2018.

[2] D. Eckhardt, A. Leiras, and A. M. T. Thomé, "Systematic literature review of methodologies for assessing the costs of disasters," *International Journal of Disaster Risk Reduction*, vol. 33, pp. 398–416, 2019.

[3] J. Gallego, "Natural Disasters and Clientelism: The Case of Floods and Landslides in Colombia," *Electoral Studies*, vol. 18, pp. 1–50, 2018.

[4] X. Wu, Z. Wang, G. Gao, J. Guo, and P. Xue, "Disaster probability, optimal government expenditure for disaster prevention and mitigation, and expected economic growth," *Science of the Total Environment*, vol. 709, p. 135888, 2020.

[5] P. Li *et al.*, "Prediction of heavy precipitation in the eastern China flooding events of 2016: Added value of convection-permitting simulations," *Quarterly Journal of the Royal Meteorological Society*, vol. 145, no. 724, pp. 3300–3319, 2019.

[6] Badan Nasional Penanggulangan Bencana, "Bencana Alam di Indonesia Menurut Waktu Tahun 2010 s/d 2019," *Badan Nasional Penanggulangan Bencana*, 2019. [Online]. Available: http://bnpb.cloud/dibi/Table3a.

[7] P. Parulian *et al.*, "Analysis of Sequential Order Incremental Methods in Predicting the Number of Tolls Affected by Disasters," *Journal of Physics: Conference Series*, vol. 1255, no. 1, pp. 1–6, 2019.

[8] A. Wanto, M. Zarlis, Sawaluddin, and D. Hartama, "Analysis of Artificial Neural Network Backpropagation Using Conjugate Gradient Fletcher Reeves in the Predicting Process," *Journal of Physics: Conference Series*, vol. 930, no. 1, pp. 1–7, 2017.

[9] M. K. Z. Sormin, P. Sihombing, A. Amalia, A. Wanto, D. Hartama, and D. M. Chan, "Predictions of World Population Life Expectancy Using Cyclical Order Weight / Bias," *Journal of Physics: Conference Series*, vol. 1255, no. 1, pp. 1–6, 2019.

[10] T. Afriliansyah *et al.*, "Implementation of Bayesian Regulation Algorithm for Estimation of Production Index Level Micro and Small Industry," *Journal of Physics: Conference Series*, vol. 1255, no. 1, pp. 1–6, 2019.

[11] G. W. Bhawika *et al.*, "Implementation of ANN for Predicting the Percentage of Illiteracy in Indonesia by Age Group," *Journal of Physics: Conference Series*, vol. 1255, no. 1, pp. 1–6, 2019.

[12] I. S. Purba *et al.*, "Accuracy Level of Backpropagation Algorithm to Predict Livestock Population of Simalungun Regency in Indonesia Accuracy Level of Backpropagation Algorithm to Predict Livestock Population of Simalungun Regency in Indonesia," *Journal of Physics: Conference Series*, vol. 1255, no. 1, pp. 1–6, 2019.
[13] N. L. S. R. Ginantra *et al.*, "Architectural Models for Predicting the Total of Natural Disasters and their Effects Using Batch Training," *Journal of Physics: Conference Series*, vol. 1566, no. 1, p. 012032, 2020.
[14] A. Wanto *et al.*, "Analysis of the Accuracy Batch Training Method in Viewing Indonesian Fisheries Cultivation Company Development," *Journal of Physics: Conference Series*, vol. 1255, no. 1, pp. 1–6, 2019.
[15] B. Febriadi, Z. Zamzami, Y. Yunefri, and A. Wanto, "Bipolar function in backpropagation algorithm in predicting Indonesia's coal exports by major destination countries," *IOP Conference Series: Materials Science and Engineering*, vol. 420, no. 012087, pp. 1–9, 2018.
[16] N. Nasution, A. Zamsuri, L. Lisnawita, and A. Wanto, "Polak-Ribiere updates analysis with binary and linear function in determining coffee exports in Indonesia," *IOP Conference Series: Materials Science and Engineering*, vol. 420, no. 012089, pp. 1–9, 2018.
[17] A. Wanto and J. T. Hardinata, "Estimations of Indonesian poor people as poverty reduction efforts facing industrial revolution 4.0," *IOP Conference Series: Materials Science and Engineering*, vol. 725, no. 1, pp. 1–8, 2020.
[18] A. Wanto *et al.*, "Model of Artificial Neural Networks in Predictions of Corn Productivity in an Effort to Overcome Imports in Indonesia," *Journal of Physics: Conference Series*, vol. 1339, no. 1, pp. 1–6, 2019.

IOP Conf. Series: Materials Science and Engineering 1071 (2021) 012023 doi:10.1088/1757-899X/1071/1/012023

Decision Making Analysis for Free Internet Quota Assistance Online Learning during the Covid-19 Pandemic

E Budiman[1*] **and U Hairah**[1]
[1]Department of Informatics, Engineering Faculty, Universitas Mulawarman, Samarinda, East-Kalimantan, Indonesia

E-mail: edy.budiman@fkti.unmul.ac.id

Abstract. One of the Indonesian government's efforts to reduce the impact of the crisis due to the Covid-19 pandemic is through the distribution of social assistance through the distribution of free internet data for students to support online learning during the implementation of government policies regarding large-scale social restrictions. This social assistance involves 350 students as an alternative as a target for decision making. In an effort to support this policy, this study offers a multi-criteria decision-making analysis approach using the Moora and Vikor methods. The results of this study present the optimal performance of the accuracy, precision and error rate of the two methods.

1. Introduction

In order to reduce the spread of the Covid-19. In Indonesia, not through quarantine or regional lockdowns, but through large-scale social restrictions. Large-scale social restrictions are restrictions on activities to prevent the spread of disease in areas suspected of being infected or contaminated with the disease. This policy covers a number of points, including school and work holidays, to limiting activities in public places. Government Regulation on Large-Scale Social Restrictions[1] for the handling of the Covid-19 pandemic.

In supporting government policies to implement social restrictions, the role of information and communication technology (ICT) is very large to support the use of learning media platforms such as teleconferencing, telemedicine and e-learning. The use of this ICT-based platform eases the need to work, learn and even pray without having to leave the house. However, some groups still face challenges such as network availability and affordability of internet data usage, particularly for students in accessing distance learning facilities with limited budgets(cost). For this reason, the Indonesian Government's policy is to re-implement a Regulation through the Secretary of the Ministry of Education and Culture concerning Technical Guidelines for Internet Data Quota Assistance in 2020[2]. These technical guidelines serve as guidelines in distributing internet data quota assistance for educators and students so that they can support the implementation of distance learning during the COVID-19 pandemic. Internet data quota assistance is provided to students and teachers or lecturers. The technical guidelines explained that the form of assistance provided was in the form of internet data quota with details divided into general quota and learning quota.

The internet quota package for students in early childhood education (PAUD) gets 20 GB per month with details of 5 GB for general quota and 15 GB for learning quota. Elementary, Junior and

Senior High School students get 35 GB per month, with details of 5 GB for general quota and 30 GB for learning quota[2].

The policy of free internet assistance to higher education institutions (lecturers and students) is considered less proportional and right on target to users according to their learning needs, due to several factors such as the amount of learning load; courses and academic credit, and students' economic ability problems. For this reason, this research is expected to be able to become the basis of policy for decision makers in managing free internet assistance based on the need for internet data consumption to users in the use of internet access for the integrity of online learning.

This study aims to apply a multi-criteria decision-making method in supporting the management of distribution of free internet assistance to students, in particular, to determine the performance of the Multi Objective Optimization method on the Basic of Ratio Analyzer (MOORA) and Visekriterijumsko Kompromisno Rangiranje (VIKOR) for decision making for prospective beneficiaries based on five criteria; internet quota, academic credit, number of courses, ability purchase quota and student monthly costs. To measure the performance method, the confusion matrix testing is used. The result of method preference is the actual data of beneficiaries.

The research contribution is expected to be a reference for decision support analysis approaches in the management of internet data package assistance for online learning needs during the Covid-19 pandemic in education.

2. Testing Methodology and Analysis

An overview of the analytical methodology of decision making and performance testing of multi-criteria methods is presented in **Figure 1.**

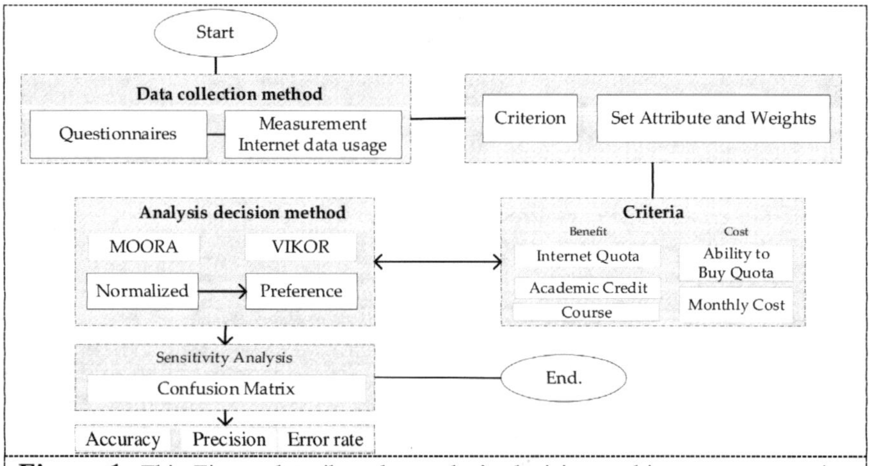

Figure 1. This Figure describes the analysis decision making process starting from data collection to the performance testing of the method.

2.1. Data Collection Methods

Data collection through observation and measurement of internet data usage during online learning, data is taken from the undergraduate program in informatics department Mulawarman University, East Kalimantan, Indonesia, where the population is all students who are currently carrying out online learning from home during the Covid-19. From the results of observations, a total sample of 350 students filled out the questionnaire. Questionnaire techniques are used to obtain data criteria; the number of courses, academic credits and the student's economic ability[3], while the measurement is used to find out how much internet data package usagein each meeting (duration 45 minutes).

2.2. Criteria and Attribute Set

Determination of the criteria for decision making originates from observation and measurement of data consumption, and the criteria data are obtained along with the weights of importance, including the benefit or cost attributes. The criteria and attributes are shown in Table 1.

Table 1. Criteria, weighting and attributes.

Criteria	**Crisp**	**Weights**	**Attribute**
C1: Internet Quota	Internet data usage (MB)	5→(0.333)	Benefit
C2: Academic Credit	14 to 24	4→(0.267)	Benefit
C3: Amount Course	5; 6; 7; 8 ; 9	3→(0.2)	Benefit
C4: Quota Purchase Ability	IDR 100.000 up to IDR 400.000	2→(0.133)	Cost
C5: Monthly Cost	IDR 500.000 up to IDR 2.000.000	1→(0.067)	Cost

The weighting importance method in Table 1 criteria used the Rank Sum (RS) ranking technique which refers to Mats D and L Ekenberg[4],[5] where the weights for the 5 criteria for each straight rank (rj) value are C1 =0.333, C2 =0.267, C3 =0.2, C4 = 0.133 and C5 =0.067.

2.3. Decision Analysis Methods

Moora method, for the analysis of work stages this method refers to W Brauers et al [6],[7] with a calculation analysis such as in the following equation:

$$x_{ij}^{*} = \frac{x_{ij}}{\sqrt{\sum_{j=1}^{m} x_{ij}^{2}}}, \quad y_{j}^{*} = \sum_{i=1}^{i=g} x_{ij}^{*} - \sum_{i=g+1}^{i=n} x_{ij}^{*}, \tag{1}$$

Vikor method, the analysis stage this method according to A Mardani et al[8], analysis calculation as in the following equation:

$$Q_1 = v\left[\frac{(s_i - s^{max})}{(s^{max} - s^{min})}\right] + (1 - v)\left[\frac{(r_i - r^{max})}{(r^{max} - r^{min})}\right] \tag{2}$$

2.4. Performance Decision Analysis Methods

The confusion matrix represents the performance and actual conditions of the data for each preference methods. will test the method; accuracy, precision, and error rate. The calculation for each test following equation[9],[10] in table 2.

3. Results

The analysis results of the Moora method use equation (1) through several process; weighting calculation, normalized data and preferences. Briefly, the calculation results of the preference value of 350 alternatives along with their attributes (benefit-cost) and criteria are presented in Table 4 and Figure 2 and 3.

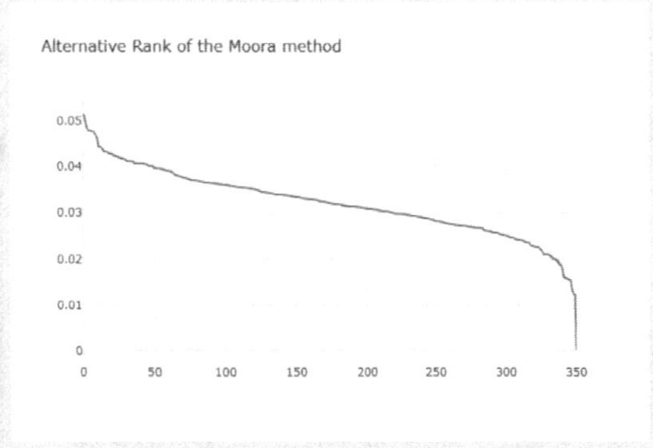

Figure 2. Alternative rank of the Moora

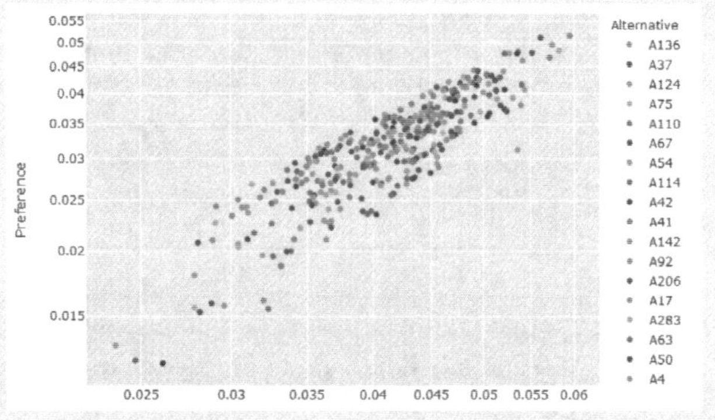

Figure 3. Scatter of the Moora references

Table 4. Reference result of Moora analysis method

Alt.	Internet Quota	Academic Credit	Courses	Quota Purchase Ability	Monthly Cost	Benefit	Cost	Preference
A1	456.640	24	9	100000	700000	0.04328	0.00503	0.0383
A2	582.180	24	9	200000	800000	0.04741	0.00731	0.0401
A3	530.510	24	9	400000	2000000	0.04571	0.01679	0.0289
A↓	⋮	⋮	⋮	⋮	⋮	⋮	⋮	⋮
A349	665.400	20	5	300000	1000000	0.04444	0.00847	0.0359
A350	614.100	20	8	200000	1200000	0.05360	0.02237	0.0312

Furthermore, the preference value of each alternative is sorted from largest to smallest (descending), so that the result of the ranking of preference values is the alternative ranking order. The results of sorting preference values are presented in the Table 5.

Table 5. Ranking result of Moora analysis method

Alt.	Preference	Sorting	Alt.	Ranking
A1	0.0383		A136	0.0516
A2	0.0401		A37	0.0512
A3	0.0289		A124	0.0496
A4	0.0431		A75	0.0484
A5	0.0343		A110	0.0481
A↓	⋮		A67	0.0479
A346	0.0349		A54	0.0478
A347	0.0336		A114	0.0478
A348	0.031		A42	0.0477
A349	0.0359		A41	0.0469
A350	0.0312		A142	0.0463

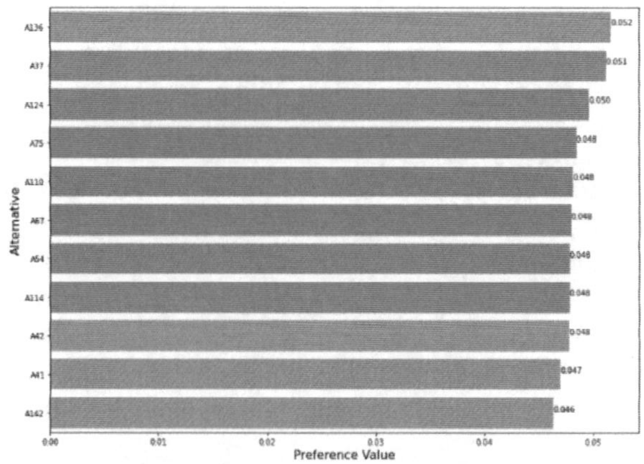

Figure 4. Alternative Rank of the Moora method

IOP Conf. Series: Materials Science and Engineering 1071 (2021) 012023 doi:10.1088/1757-899X/1071/1/012023

3.1. Result: Vikor analysis method

The analysis Vikor results method using equation (2) are presented in Table 6 and Figure 5 and 6.

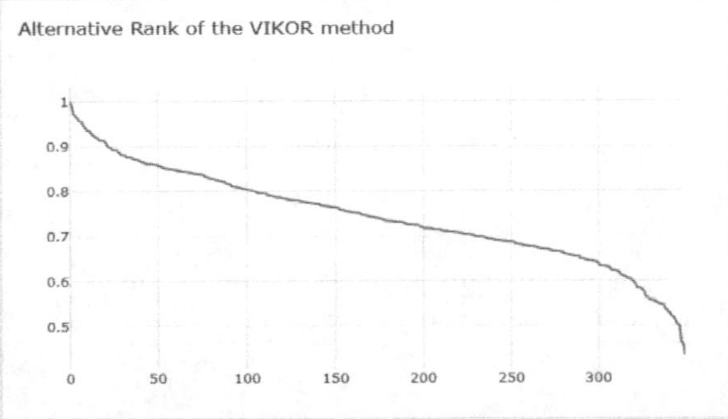

Figure 5. Alternative rank of the Vikor

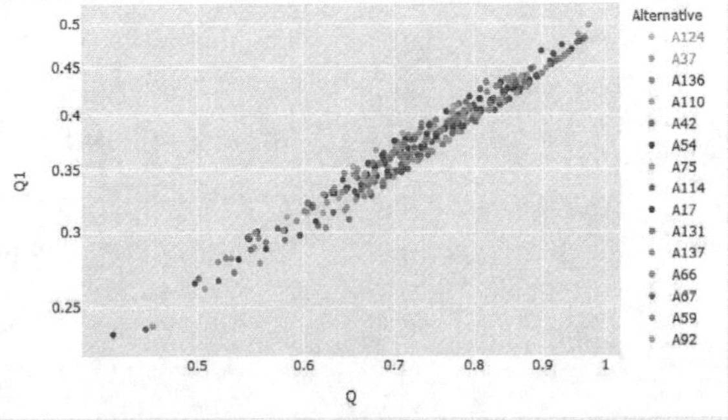

Figure 6. Scatter of the Vikor references

Table 6. Reference result of Vikor analysis methods

Alt.	Internet Quota	Academic Credit	Courses	Quota Ability	Monthly Cost	Q1	Q2	Q
A1	456.640	24	9	100000	700000	0.4245	0.4386	0.8631
A2	582.180	24	9	200000	800000	0.4296	0.4225	0.852
A3	530.510	24	9	400000	2000000	0.3766	0.3472	0.7237
A↓	⋮	⋮	⋮	⋮	⋮	⋮	⋮	⋮
A349	665.400	20	5	300000	1000000	0.3971	0.3893	0.7864
A350	614.100	20	8	200000	1200000	0.4270	0.4407	0.8676

Furthermore, the preference value of each alternative is sorted from largest to smallest (descending), so that the result of the ranking of preference values is the alternative ranking order. The results of sorting preference values are presented in the Table 7.

Table 7. Ranking result of Vikor analysis method

Alt.	Preference	Sort	Alt.	Ranking
A1	0.8631		A124	1
A2	0.852		A37	0.9929
A3	0.7237		A136	0.9728
A4	0.9117		A110	0.967
A5	0.7505		A42	0.9648
A↓	⋮		⋮	⋮
A346	0.7061		A164	0.5012
A347	0.7005		A184	0.4979
A348	0.6712		A215	0.4641
A349	0.7864		A174	0.4586
A350	0.8676		A299	0.4346

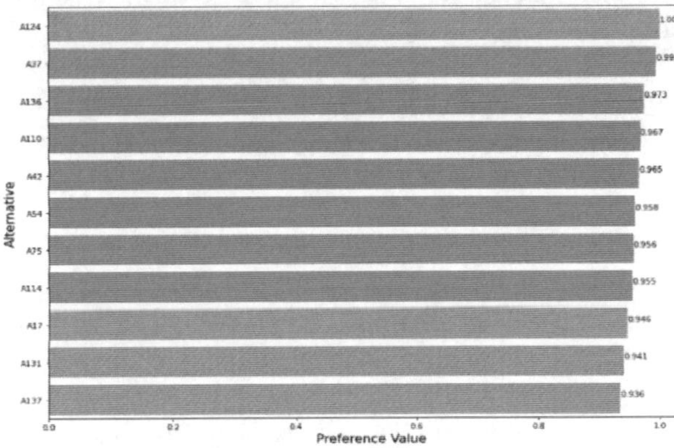

Figure 7. Alternative Rank of the Vikor method

3.2. Performance analysis methods
Performance testing uses a confusion matrix that compares/matches the ranking results of the Moora and Vikor methods toward actual data shown in Table 8.

Table 8. Performance of Moora and Vikor method toward actual data

	Moora method			Vikor method		
Alt.	Accuracy	Precision	Error rate	Accuracy	Precision	Error rate
10	0.954	0.2857	0.040	0.951	0.2308	0.049
20	0.917	0.3548	0.083	0.906	0.2593	0.094
30	0.874	0.3478	0.126	0.863	0.2857	0.137

4. Discussion
From the research results that have been presented in the previous section, several important points were found ie 1). Both methods of decision-making analysis (Moora and Vikor) have their own preferred characteristics, where the accuracy, precision and error rate of the Moora method are better than the Vikor method (for the case of five criteria; 350 alts.). Another thing is that the weighting method also has an effect on the preference value, so it has a significant effect on performance results. 2). The performance results of the Moora and Vikor methods show a decrease in the level of accuracy, precision and error rate when the target beneficiary is increased, the more target beneficiaries, the lower the performance of the method. The importance of criteria in terms of how much weight is given to them in the decision-making process. In determining the decision-making method for the case of social assistance management according to the needs and targets, this affects the value of accuracy and performance accuracy. Free internet data assistance for students is a case study of how the performance of decision-making analysis methods is applied and the results of this study have provided interesting things in the future for further studies[11].

5. Conclusions
The importance of criteria in terms of how much weight is given to them in the decision-making process. In determining the decision-making method for the case of social assistance management according to the needs and targets, this affects the value of accuracy and performance accuracy. Free internet data assistance for students is an example of a case study on how the performance of decision-making methods is applied, and the results of this study have provided interesting things in the future for further studies. This study offers a multi-criteria decision-making analysis approach using the Moora and Vikor methods. The research results of both methods produce the best alternative preference from a set of targeted alternatives. The results of this study still require better performance optimization through the development of criteria, weighting methods, data normalization and other decision-making methods, considering that this social assistance case study is very important because it is a concern for the welfare of the community[12].

References
[1] Peraturan Pemerintah Republik Indonesia 2020 Peraturan Pemerintah Nomor 21 Tahun 2020 tentang Pembatasan Sosial Berskala Besar Dalam Rangka Percepatan Penanganan Coronavirus Disease 2019
[2] Kemendikbud 2020 *Peraturan Sekretaris Jenderal Nomor 14 Tahun 2020, tentang Petunjuk Teknis Bantuan Kuota Data Internet Tahun 2020* (Jakarta)
[3] Budiman E 2020 Mobile Data Usage on Online Learning during COVID-19 Pandemic in Higher Education *International Journal of Interactive Mobile Technologies (iJIM)* **14**
[4] Danielson M and Ekenberg L 2017 Trade-offs for ordinal ranking methods in multi criteria

decisions *Lecture Notes in Business Information Processing*

[5] Budiman E 2020 Importance-weighted Ranking Methods for Preference the Covid-19 Pandemic Social Assistance *International Journal of Engineering and Advanced Technology (IJEAT)* **10** 108–15

[6] Brauers W K M and Zavadskas E K 2012 Robustness of MULTIMOORA: A method for multi-objective optimization *Informatica*

[7] Pakpahan H S, Anandiya O, Hairah U and Wati M 2019 Decision support system for predicting increased data on objects of motor vehicle name transfer (BBNKB I) using trend moment method (Case study: Wheels 2 and wheels 4 in Samarinda) *Journal of Physics: Conference Series*

[8] Mardani A, Zavadskas E K, Govindan K, Senin A A and Jusoh A 2016 VIKOR technique: A systematic review of the state of the art literature on methodologies and applications *Sustainability (Switzerland)*

[9] Visa Sofia D 2011 Confusion Matrix-based Feature Selection Sofia Visa *ConfusionMatrix-based Feature Selection Sofia*

[10] Budiman E, Haviluddin, Dengan N, Kridalaksana A H, Wati M and Purnawansyah 2018 Performance of Decision Tree C4.5 Algorithm in Student Academic Evaluation *Lecture Notes in Electrical Engineering* (Springer Verlag.) pp 380–389

[11] Wati M, Novirasari N, Budiman E and Haeruddin 2018 Multi-Criteria Decision-Making for Evaluation of Student Academic Performance Based on Objective Weights *2018 Third International Conference on Informatics and Computing (ICIC)* (Palembang: IEEE) pp 1–5

[12] Budiman E, Dengen N, Haviluddin and Indrawan W 2017 Integrated multi criteria decision making for a destitute problem *2017 3rd International Conference on Science in Information Technology (ICSITech)* (Bandung: IEEE) pp 342–7

Analysis of Custom IoT Based Electricity Monitoring System

D Eridani[1]*, E D Widianto[1] and A R Yulianto[1]
[1]Dept. of Computer Engineering Diponegoro University Semarang

E-mail: dania@ce.undip.ac.id

Abstract. Electricity is a resource that currently exists in various sectors of the community. Each person's electricity needs are different depending on the needs of the electronic equipment used. Excessive use of electronic equipment creates extreme electrical power. Electricity consumption is an issue that discussed since a long ago. This research aims to present a custom electronic sensor to monitor electricity consumption based on the Internet of Things. It also explained the correlation between input and output points from the custom electronic sensor. This research method consists of 4 steps: system requirement, system design, system implementation, and last, testing and analysis. The system developed is able to get the current, voltage, and power data. The data then processed by Arduino UNO and directly send by ESP8266 to Blynk Application on Android smartphones via Internet communication. The voltage testing is done in the voltage and current sensors to get the program's main loop's constant values. The system input value and the Blynk Application value are also measured to check the system's error.

1. Introduction

Electricity is a resource that currently exists in various sectors of the community. Each person's electricity needs are different depending on the needs of the electronic equipment used. Excessive use of electronic equipment creates extreme electrical power. Electricity consumption is an issue that discussed since a long ago. The energy impact also has been carried out since it generally indicates the environmental effects and the direct impact of ICT equipment on electricity consumption [1]. Research about the trend of electricity usage also done to check the appliances usage behavior [1]–[5] Besides checking the electricity consumption, it is also essential to research electricity saving. Research provides a comparison of electricity saving tariff [6] and evaluation of household electricity savings are done [7]. To prevent waste electricity usage, various research to develop an electricity monitoring system, and a technique to estimate power consumption are also done [8]–[11] Besides used as electricity monitoring usage, a monitoring system can be used to check the parameter's value such as voltage, current, and power to prevent electronic equipment damage. That's how monitoring electricity usage can be crucial.

This research aims to present a custom electronic sensor to monitor electricity consumption based on the Internet of Things. Different from [12], this research using an Android smartphone with Blynk Application instead of a web-based monitoring system. Compared to [13], this research focused on the development process and the correlation of the input and output current and voltage measured in the custom electronic sensor.

2. Literature Review

2.1. Arduino UNO, ESP 8266, and Blynk

Arduino UNO is a microcontroller board based on Atmega 328P with 14 digital I/O pin (6 pin support as PWM) and six analog inputs. It provides USB and power jack connection. USB connector, AC/DC adaptor, and batteries can be used to supply the board [14]. It is the board to start with, and it comes in two flavors, through-hole and SMD. Compared to Arduino Mega, Arduino Due, and Arduino Leonardo, it has the correct amount of pin to be used in the system [15]. Arduino UNO does not come with wireless communication so that it is usually added with another board to support communication via the Internet.

Comparing to several wireless boards named Arduino Wi-Fi Shield, Genuino Yun Shield, Genuino MKR 1000, ESP8266, and Onion Omega, ESP8266 is suitable for this system. The board using Tensilica Xtensa LX106 processor, support 5 V from USB or 3.3 V from VIN power supply, 3.3 V I/O, Internal RAM: 64 KB for instructions, Internal RAM: 96 KB for data, QSPI flash: 512 KB to 4 MB. It also supports SPI, I2C, I2S, and UART. It also supports the Arduino IDE and has the smallest board dimension and price compare to others [16].

Blynk was designed for the Internet of Things. It supports mobile IoS and mobile Android to control Arduino, Raspberry Pi, and other hardware via the Internet. It provides three major components which are the Blynk App, Blynk Library, and Blynk Server [17].

2.2. Custom Sensor Circuit

The custom sensor circuit used consists of several parts, the voltage sensor circuit, current sensor circuit, voltage regulator circuit, and Operational Amplifier circuit in the current sensor, as shown in Figure 1.

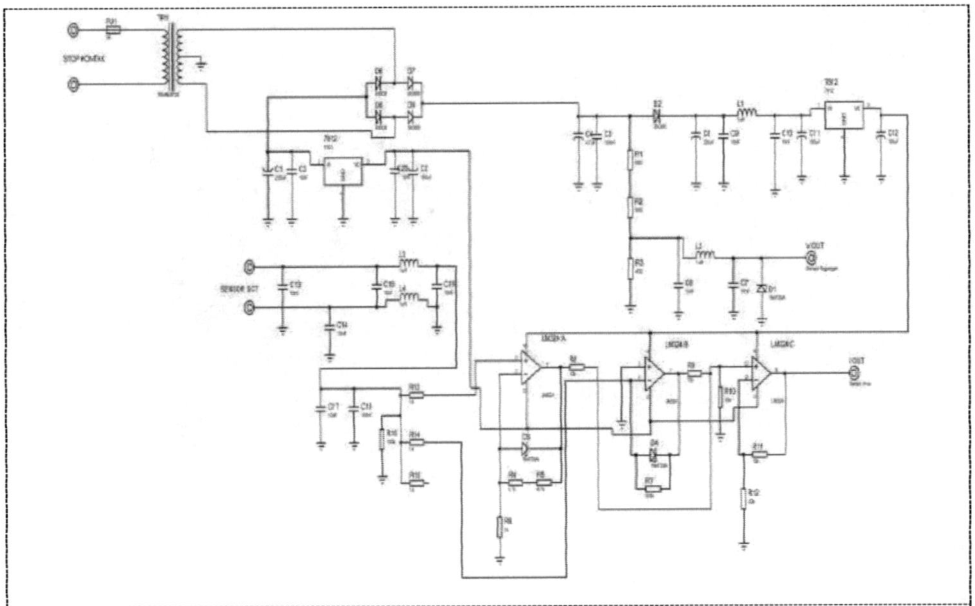

Figure 1. Custom Electronic Sensor Circuit Design

The voltage sensor circuit is used to lower the voltage from the resource to Arduino UNO. It uses CT 18 V step-down transformer to reduce the maximum 220 V to a maximum 36 V. The output is connected to Diode Bridge to change the AC to DC. The current sensor circuit use to lower the voltage ripple that goes to SCT 013-010 sensor. The voltage regulator circuit gives stable 12 V to IC

LM324 [18] in the Op-Amp circuit. The Op-Amp circuit itself uses to reinforce the input from the current sensor circuit so that it can go to Arduino UNO [19].

3. Research Method

The method of this research consists of 4 steps. The first step is system requirement analysis, then system design, system implementation, and last are testing and analysis. From the application requirement analysis, the component's list and the application's functional needs are found. The operational needs of the system are:

- The system must be able to communicate with the Blynk application in an Android smartphone using ESP8266 with Wi-Fi.
- The system can read and process the value from the current and voltage sensor in the system.
- The system can calculate real electricity power used from the current sensor and voltage value.
- The system can show the current value, voltage value, and power value on Blynk Application in Android smartphones.

The non-function needs of the system are:

- The system using the Arduino UNO board as the central controller to process the data, and ESP8266 to send the data to Android smartphones.
- The system using the SCT-013-010 sensor to check the current value in each plug. It uses CT 18V step-down transformer to proceed with the voltage sensor circuit to check each plug's voltage value.
- ESP8266 board used to send the data from Arduino UNO to the Blynk application.
- Arduino UNO in the system using a 9V power supply.

The next step of the research is system design. The system design can be seen in Figure 2. As mentioned before, the sensor circuit used in this research consists of several parts, namely the voltage sensor circuit, the current sensor circuit, the voltage regulator circuit, and the Operational Amplifier circuit.

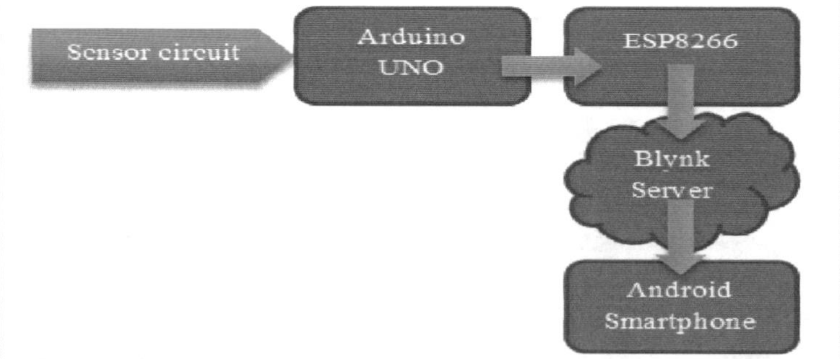

Figure 2. System Design.

The current sensor circuit used to make large input voltage from the voltage source readable in Arduino UNO. The current sensor circuit is used to reduce the voltage ripple at the 013-010 SCT sensor input. The voltage regulator circuit provides a stable voltage of 12V on the LM324 IC in the Operational Amplifier circuit. The Operational Amplifier circuit serves to amplify the input from the current sensor circuit so that it can be accepted on the Arduino UNO.

As for the Android Smartphone, the interface used is from the Blynk Application. Several widgets can be used in Blynk, as on this system used voltage sensor display, current sensor display, power display, and super chart. Super chart widget is used to display the current, voltage, and power result sent from the ESP8266. It is also set to read until two digits decimals.

4. Testing and Analysis

The testing is done by first checking each voltage value read in each input sensor and output. Second, by checking the comparison value between the system and the Blynk application display as a whole system. The test was carried out using a digital multimeter. The load used for testing comes from 1 electric iron, two mobile phone chargers, and one laptop charger.

4.1. Voltage Sensor Circuit Test

The first test is the voltage sensor circuit test. The voltage sensor circuit's voltage value is used to compare the actual voltage input value from the voltage sensor circuit and the real output value from the voltage sensor circuit. The value of the input voltage and output voltage used in the calculation of the voltage sensor will applied in the void loop function on the Arduino UNO board. Testing is done 50 times by measuring the voltage value at the voltage source, then the 18V 500 mA transformer CT's output value. It will then continue with measuring the diode bridge's output value where the value has changed from analog to digital. And the last is measuring the voltage sensor circuit's output value, which is used as a benchmark to determine the ratio of the output voltage value between the sensor circuit and the voltage source. The voltage sensor testing can be seen in Figure 3.

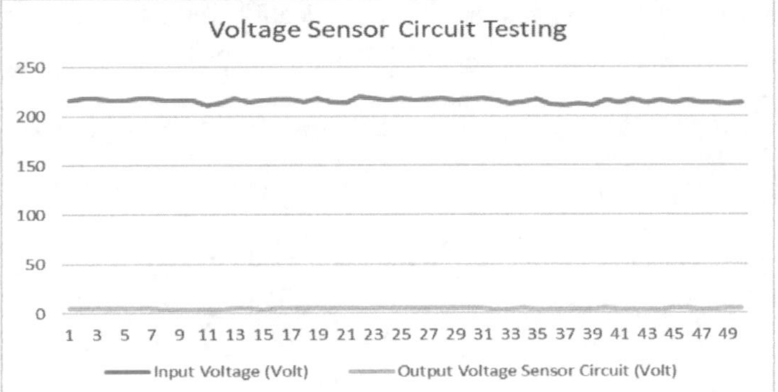

Figure 3. The voltage sensor circuit testing.

The result from Figure 3 showed the comparison of the 50 value of the voltage sensor circuit, CT transformer output voltage, diode bridge output voltage, and output voltage sensor circuit. All 50 values from the input voltage were added and divided with all of the output voltage from the voltage sensor circuit. The result is 47.6 V. The value result is then used as the constant in the voltage calculation in the Arduino IDE void loop, as shown in Blynk Application.

4.2. Current Sensor Circuit Test

The current sensor circuit testing is used to get the comparison of actual voltage value from SCT 013-10 sensor input and current sensor circuit output. The comparison result of the voltage value is used as the current calculation in the Arduino IDE void loop. The SCT sensor has an output where every 10 Ampere read is the same as 1 V sensor output. The testing was done by measuring the output voltage on SCT 013-010 sensor output, the output voltage on OP-amp 1, the output voltage on OP-amp 2, the output voltage on OP-amp 3. The testing is also done in 50 times to check the reinforcement value from the current circuit sensor. The current sensor testing can be seen in Figure 4.

The result of Figure 4 showed a comparison of each point in the current sensor circuit. The reinforcement value got from the sum of input value then divided with the sum of the gates' outputs. The result is 31.5 V, and this value is used as the constant in Current computation in the Arduino IDE void loop as send to Blynk Application.

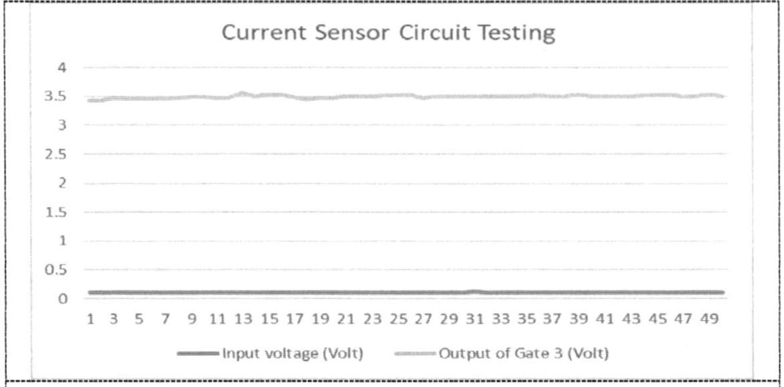

Figure 4. The current sensor circuit testing.

4.3. System Test

After each sensor circuit testing, the final testing is complete system testing. This testing done to check the differences between the values showed in Blynk Application and the current and voltage input in the system. Figure 5 and Figure 6 showed the comparison value of it.

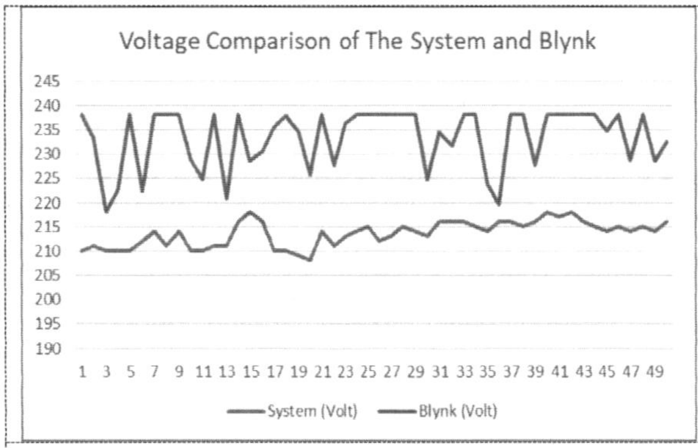

Figure 5. Voltage Comparison on The System and Blynk.

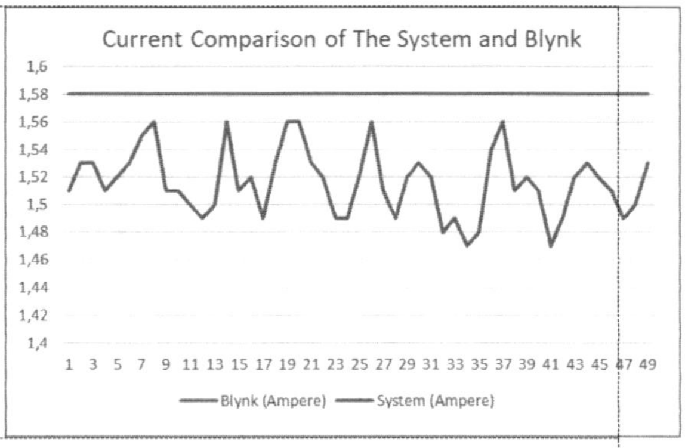

Figure 6. Current Comparison on The System and Blynk

Figure 5 showed the comparison voltage value between the system and the Blynk as a whole system. The average error voltage sensor is 8.41%. Figure 6 showed the current comparison value between the system and the Blynk as a whole system. The average error current sensor is 4.14%. Based on the result, it can be concluded that there 8.41% differences in voltage sensor and 4.14% differences in the current sensor. The possibility of the error happened because the measurement process is done using the values measured at each point on the system. It is better to measure the value by using the peak of the voltage and current value using differentiation.

5. Conclusion

It can be concluded that the system build can monitor the current and voltage used in the measuring point and send it to the Blynk application directly. There are 8.41% differences in voltage sensors and 4.14% differences in current sensors during the input and output system's measurement process. It is better to measure the value by using the peak of the voltage and current value using differentiation as a suggestion.

Acknowledgment
This research was financially supported by Faculty Engineering, Diponegoro University, Indonesia, through Strategic Research Grant 2019.

References
[1] Pothitou M Hanna R F and Chalvatzis K J, 2017 ICT entertainment appliances' impact on domestic electricity consumption *Renew. Sustain. Energy Rev.* **69**, June 2015 p. 843–853.
[2] Firth S Lomas K Wright A and Wall R, 2008 Identifying trends in the use of domestic appliances from household electricity consumption measurements *Energy Build.* **40**, 5 p. 926–936.
[3] Shiraki H Nakamura S Ashina S and Honjo K, 2016 Estimating the hourly electricity profile of Japanese households – Coupling of engineering and statistical methods *Energy* **114** p. 478–491.
[4] Ohler A M Loomis D G and Ilves K, 2020 A study of electricity savings from energy star appliances using household survey data *Energy Policy* **144**, April p. 111607.
[5] Azarova V Cohen J J Kollmann A and Reichl J, 2020 Reducing household electricity consumption during evening peak demand times: Evidence from a field experiment *Energy Policy* **144**, January p. 111657.
[6] Prasanna A Mahmoodi J Brosch T and Patel M K, 2018 Recent experiences with tariffs for saving electricity in households *Energy Policy* **115**, March 2016 p. 514–522.
[7] Laicane I Blumberga D Blumberga A and Rosa M, 2015 Evaluation of Household Electricity Savings. Analysis of Household Electricity Demand Profile and User Activities *Energy Procedia* **72** p. 285–292.
[8] Potapovs A and Obushevs A, 2017 Development and Comparison of Technical Solutions for Electricity Monitoring Equipment *Latv. J. Phys. Tech. Sci.* **54**, 6 p. 3–9.
[9] Laple R Putra S Febia T Q Ndala S and Santoso A J, 2017 Analysis of Prepaid Electricity Monitoring System with SMS Notification **9**, 12 p. 304–307.
[10] Wood G *et al.*, 2019 Sensors, sense-making and sensitivities: UK household experiences with a feedback display on energy consumption and indoor environmental conditions *Energy Res. Soc. Sci.* **55**, March 2018 p. 93–105.
[11] Lam C Zheng Z and Luo X, 2019 Electricity load decomposition prototype for household appliances: System design and development *Energy Procedia* **158** p. 3158–3163.
[12] Alulema D Zapata M and Zapata M A, 2018 An IoT-Based Remote Monitoring System for Electrical Power Consumption via Web-Application in *2018 International Conference on Information Systems and Computer Science (INCISCOS)* p. 193–197.
[13] Suwansit K Konsombut B Hankongkaew P and Tantidham T, 2014 PMA: Power Monitoring Application for Android in *2014 Third ICT International Student Project Conference (ICT-ISPC)* p. 69–72.
[14] Farnell, 2013 Arduino Uno Datasheet *Datasheets* p. 1–4.
[15] Tabular T Arduinos W and Further G, Arduino Comparison Guide a learn . sparkfun . com. [Online]. Available: https://learn.sparkfun.com/tutorials/arduino-comparison-guide/all.pdf.
[16] Plaza P Sancristobal E Carro G and Castro M, 2018 Wireless Development Boards to Connect the World **22**.
[17] 2020, How Blynk Works. [Online]. Available: https://docs.blynk.cc/#intro-how-blynk-works.
[18] Semiconductor O, 2002 LM324,LM224,LM2902 Single Supply Quad Operational Amplifiers p. 1–12.
[19] Carter B and Huelsman L P, 2001 Handbook Of Operational Amplifier Active RC Networks *Networks* October p. 1–85.

Optimised Genetic Algorithm Crossover and Mutation Stage for Vehicle Routing Problem Pick-Up and Delivery with Time Windows

M F Ibrahim[1]*, F R Nurhakiki[1], D M Utama[2] and A A Rizaki[1]
[1]Department of Logistics Engineering, Universitas Internasional Semen Indonesia, Gresik, Indonesia.
[2]Department of Industrial Engineering, University of Muhammadiyah Malang, Malang, Indonesia.

E-mail: faisalibrahim.ie@gmail.com

Abstract. Some problems happen in transportation and distribution. Vehicle Routing Problem (VRP) can be applied in some systems above. Deciding the optimal route for every vehicle will impact increasing economic interests and expected logistical planning results. This research will raise the problem of a shipping and logistics company. There are 45 branch offices with one main depot to serve a certain area that will be a transit point before all packages will be sent to the destination. The vehicle will depart from the depot to all branch offices to delivers and pick the package up at certain hours. In every route, planning should be considered to the amount of load when loading and unloading. Every vehicle has carrying capacity, and every branch office has various loading and unloading service time windows. Based on the problem's description, this research was conducted to find the optimal solution in the Vehicle Routing Problem Pick-up and Delivery with Time Windows (VRPPDTW). An optimised genetic algorithm was developed to solve these problems by adjusting the crossover and mutation stages. The result informs that the route proposed from optimised genetic algorithms is better than the company's existing route in all aspects. On the other hand, we carried out an analysis effect of the number of iterations on distance traveled, the number of penalties, and the fitness value. This algorithm can be applied in VRPPDTW and produces an optimal solution.

1. Introduction

The vehicle routing problem (VRP) attempts to find a route to deliver goods from a central depot to a geographically dispersed destination set. Many companies face this problem, which is very important in distribution and logistics because of the need to find a cost-effective route that provides high customer satisfaction [1]. During the product distribution process, it is often a problem if a product's delivery takes longer to exceed the specified delivery working hours. It is influenced by several factors, including travel time at each destination and vehicle capacity. So the solution that needs to be done and considered is calculating the route at each destination starting from the Distribution Warehouse. This route calculation states that the vehicle will visit the customer exactly once according to the shortest route's order until the vehicle returns to the depot again. The issues raised in this article will lead to Vehicle Routing Problem Pick-up and Delivery with Time Windows (VRPPDTW).

VRPPDTW has the primary goal of determining the needs of resources and routes with minimum costs, but in pick-up and delivery, consider the capacity and time windows [1]. VRPPDTW based on the integration of vehicles that deliver and pick-up goods in the transport network that is affected by time, thus enabling the optimization of the combined assignment of goods into vehicles and routing of congested transportation networks [2]. Pick-up and delivery systems refer to how the fleet fulfills the destination node's request, and the time window refers to the time when the fleet reaches the customer's node. There are two types of time windows, and it imposes that each vehicle delivers the goods to the customers within a specific time interval, and there are two types of time windows [3]. The vehicle may not arrive more than the customer's node closing time or before the customer's node open time. The best route must be on time and disciplined capacity as well as minimum distance and cost.

2. Optimised Genetic Algorithm

Holland [4] developed Genetic Algorithms (GA) and continued to use it to solve complex problems. When the optimal solution fails to find another search, GA is often tried to solve the problem. Several previous studies, including Whitley, Starkweather [5], Sponsler [6], and Marin, Gonzalez [7], apply GA to solve complex problems when optimal solutions fail to be found by another search. GA can solve complex problems with relatively shorter computation time to reach near-optimal solutions. Recent research that uses genetic algorithms in complex Vehicle Routing Problems (VRP) problems have been carried out by Mohammed, Abd Ghani [8], de Oliveira da Costa, Mauceri [9], Liu and Jiang [10], and Sethanan and Jamrus [11]. Complex multi-constraint problems were successfully solved in their study, and came up with a suitable solution. According to Ibrahim, Masudin [12], Mohammed, Abd Ghani [8], Ho, Ho [13] and de Oliveira da Costa, Mauceri [9], GA is very efficient for solving problems with high complexity such as VRP.

An optimised genetic algorithm is applied in solving the problem of determining the vehicle route of a shipping and logistics company. This problem deserves to be a real example of the VRPPDTW application. The flow chart optimised genetic algorithm used can be seen in Figure 1.

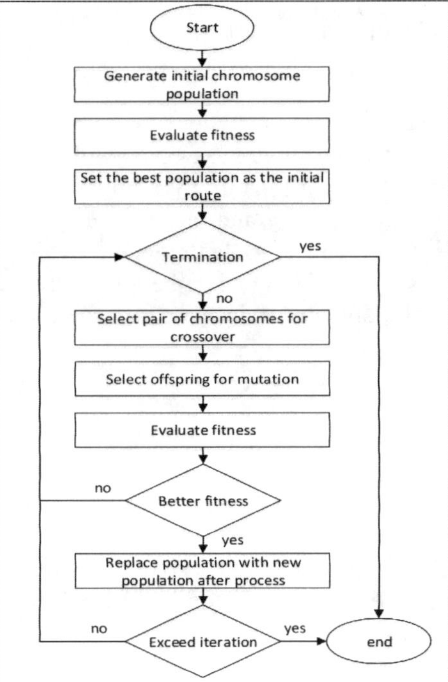

Figure 1. Optimised genetic algorithm flow chart.

The implementation of an optimised genetic algorithm in VRPTW problems can be seen in the following algorithm:

Algorithm 1: The CVRPPDTW model
1. start;
2. population (a set of the route) initialization, use randomized method and nearest neighbor algorithm to form chromosomes (routes) structure;
3. evaluate the total length, total penalty, and fitness value of the initial population, fitness = $(1/(1 + \text{total penalty})) + (1/\text{ length of route})$;
4. choose the population with the best fitness value as the initial route;
5. loop the following steps for the specified number of iterations (the more iterations, the better):
5.1. copy chromosome structure of the best population to new population (route);
5.2. generate random numbers (R_c) in the range 0.00 to 1.00, will be compared to crossover probability (P_c);
5.3. if $R_c > P_c$ proceed to stage 5.4; if $R_c \leq P_c$ do chromosome crossover;
5.4. generate random numbers (R_m) in the range 0.00 to 1.00, will be compared to mutation probability (P_m);
5.5. if $R_m > P_m$ proceed to stage 5.6; if $R_m \leq P_m$ do chromosome mutation;
5.6. evaluate the fitness value of the result population;
5.6.1 If the fitness value is better than the initial population;
5.6.2.1 Replace population with population after the crossover and mutation process
5.6.2 If the fitness value is not better than the initial population;
5.6.2.2 Do not replace the population / Keep using the initial population
6. end of the loop;
7. write out the surviving population (best route);
End.

2.1. Establishment of Initial Population
Generally, the initial population is formed randomly, but the algorithm developed in this study uses two random and heuristic methods. According to Ibrahim, Masudin [12], Nearest Neighbor is a heuristic method that can be used quite simply and produces a good initial population. Each point is represented by one number, and then it can be called a gene. A valid single route will be represented by each chromosome that contains a collection of numbers or genes. The use of heuristic methods in genetic algorithms to form initial populations is expected to minimize the time to find the best solution.

2.2. Fitness
The fitness value in VRPPDTW cannot be directly equated with the basic VRP model. In general, the fitness value can be calculated using the total distance traveled or transportation costs. However, in this developed algorithm, the fitness value will depend not only on distance but also on capacity penalties and time windows. Both types of penalties must be considered to maintain the quality of the route in terms of pick-up and delivery cases and time windows. The fitness value will be influenced by arrival time's suitability with branch office time windows and vehicle load discipline with its capacity. Therefore, these three components must be considered in determining the fitness value. The fitness value is formulated as follows:

$$fitness = \frac{1}{1 + total penalty} + \frac{1}{length of route}$$

2.3. Selection
The selection process is carried out at this stage. Gene selection is carried out on pre-formed chromosomes. The Roulette Wheel Selection (RWS) selection method is the most commonly used method. In this method, the parents obtained based on the proportion of the chromosome's fitness value, meaning that the chromosome with the best fitness value has elected a higher likelihood than other chromosomes. Chromosome parents elected at the next selection process enter the crossover process to produce offspring chromosomes.

2.4. Optimised Crossover
Not all chromosomes from parents are chosen for crossover. The crossover probability will determine whether the crossover is done or not. We developed an optimized crossover for the operator to produce better offspring and match expectations. In this section, we will explain the optimized crossover steps that are developed to produce new offspring. The distance between genes considered in the crossover stage to create a better new population, loop the following steps as much as the number of chromosomes:

Step 1: Specify a random crossover point C_p by generating random numbers in range 1 to the value of the number of genes (node) in a chromosome;

Step 2: Find the distance between the crossover point node from the previous node $C_d = distance\ gen\ C_p - 1$ to $gen\ C_p$

Step 3: Determine the chromosomes that will be crossover by generating random numbers in the range of 1 to the value of the number of chromosomes except for the previously chosen chromosome;

Step 4: Find the distance between the crossover point node to all nodes on the chromosome selected for a crossover;

Step 5: Crossover the C_p genes with genes that have the closest distance;

2.5. Optimised Mutation
Similar to crossover, the mutation probability will influence whether the mutation process is carried out or not. In this study, the mutation process will consider the distance between genes. The hope is that considering the distance will produce a new population that has better fitness value. Apart from that, it can also shorten the computation time. Loop the following steps as much as the number of chromosomes:

Step 1: Specify a random mutation point M_p by generating random numbers in range 1 to the value of the number of genes (node) in a chromosome;

Step 2: Find the distance between the mutation point node from the previous node $M_d = distance\ gen\ M_p - 1$ to $gen\ M_p$

Step 3: Find the distance between the node $M_p - 1$ to all other nodes on the same chromosome;

Step 4: Mutation the M_p genes with genes that have the closest distance;

2.6. Establishment of a New Population
The new population, which was formed using a genetic algorithm, has previously calculated the fitness value. The new population will replace the old population if it has better fitness value. If the new population replaces the old population, all chromosomes begin in a generation replaced by chromosome results obtained after the crossover and mutation stages. On the other hand, if the resulting new one has a lower fitness value, it will not be used, and the algorithm will be repeated in the old population.

2.7. Repetition Number of Iterations
The selection process, evaluation of fitness, crossover, mutation, and new population formation is repeated until the number of iterations is reached to get the best results. The only chromosomes that

can survive are those with the best fitness value among other chromosomes. This process follows the principle of natural selection, "who is strong (has a better fitness value) will survive."

2.8. Determining the Best Solution
By comparing the fitness values obtained for each iteration, the best solution is the chromosome with the best fitness value that survives at the end of the iteration.

3. Result and Discussion
We compare the total cost of the existing route and the proposed route obtained with an optimized genetic algorithm through numerical experiments. The cost component used as a reference for calculation in this case study is the cost of fuel related to the total distance traveled from the vehicle. Table 1 presents data related to the total cost of each vehicle on the existing and proposed routes. The results of numerical calculations show that the total cost generated from the proposed route is less than the existing route. The proposed route's total cost is Rp. 196,619.09, and the total cost of the existing route is Rp. 247,649.55. It can be concluded that the route formed by an optimized genetic algorithm has succeeded in producing a route that is closer to optimal. In calculating the total cost, we have not considered the costs incurred because the vehicle arrives outside the destination time windows node and costs due to overload. However, we will compare the total penalties incurred where this penalty is related to the violation of time windows for each destination node and vehicle capacity.

Table 1. Total cost comparison Actual Route & Improved Hybrid Genetic Algorithm

Vehicle	Existing Route	Improved Genetic Algorithm Route
1	56,032.27	45,790.45
2	57,818.64	54,126.82
3	66,155.00	55,615.45
4	67,643.64	41,086.36
Total	247,649.55	196,619.09

The VRPPDTW problem in this paper uses the type of soft time windows. Penalties apply if the vehicle arrives outside the available time window (penalty time windows). Likewise, the capacity penalty applies if vehicles carry more than their capacity. The more penalties given on a route illustrate, the worse the route. We continue to analyze in-depth the penalties that occur on the proposed and existing routes.

Table 2. Fitness value Comparison of Actual Route & Improved Hybrid Genetic Algorithm

Item	Existing Route	Improved Genetic Algorithm Route
Penalty capacity	2	0
Penalty time windows (opening)	0	0
Penalty time windows (closing)	4	0
Total Penalty	6	0

Table 2 shows the comparison of the total penalty value of the proposed route and the existing route. The numerical calculation results show that there are two penalty capacity on the existing route. That means that there are two times the excess capacity of the vehicles on the existing route. This excess capacity occurs because the route is not optimal. For example, a fully-loaded vehicle visits a destination node that has fewer delivery requests than pick-up requests. If this happens in the real system, the vehicle will not carry the pick-up request because the capacity is insufficient. So it is crucial to pay attention to the number of pick-ups and deliveries, not just the order of delivery with a minimum distance.

On the other hand, there are four penalty time windows (closing) on the existing route. This can be interpreted as an event that the vehicle arrives at the destination node when the destination node has closed or the service time has ended. In VRPPDTW problems, the vehicle must arrive between the time windows (opening) and time windows (closing). The destination node cannot serve vehicles that arrive too fast or are late, so they have to wait or come back afterward. This resulted in economic losses and led to longer travel times. From the results of these numerical calculations, it is also seen that no penalties have occurred on the proposed routes generated by the improved genetic algorithm. Thus, it can be concluded that all vehicles always arrive at each destination node according to the time windows and never overcapacity.

4. Conclusion

In this paper, we study the vehicle routing problem pick-up and delivery with time windows. Where is related to the constraints of vehicle capacity and time windows of the destination node. Besides, vehicles not only deliver goods but also collect goods at the same time. We developed genetic algorithms with optimised crossover and mutation operators for solving VRPPDTW cases. Numerical experiments show that the substrate route is closer to optimal than the existing route. For future work, the researcher can modify the crossover and mutation stages to solve other VRP problems.

Acknowledgments

This research is supported and financed by the Ministry of Research, Technology, and Higher Education of the Republic of Indonesia under the PDP scheme.

References

[1] Lin CKY. A cooperative strategy for a vehicle routing problem with pick-up and delivery time windows. Computers & Industrial Engineering. 2008;55(4):766-82.

[2] Mahmoudi M, Zhou X. Finding optimal solutions for vehicle routing problem with pick-up and delivery services with time windows: A dynamic programming approach based on state–space–time network representations. Transportation Research Part B: Methodological. 2016;89:19-42.

[3] Elshaer R, Awad H. A taxonomic review of metaheuristic algorithms for solving the vehicle routing problem and its variants. Computers & Industrial Engineering. 2020;140:106242.

[4] Holland JH. Adaptation in natural and artificial systems: an introductory analysis with applications to biology, control, and artificial intelligence: University of Michigan Press; 1975.

[5] Whitley D, Starkweather T, Bogart C. Genetic algorithms and neural networks: optimizing connections and connectivity. Parallel Computing. 1990;14(3):347-61.

[6] Sponsler JL. Genetic algorithms applied to the scheduling of the hubble space telescope. Telematics and Informatics. 1989;6(3):181-90.

[7] Marin FJ, Gonzalez FJ, Sandoval F. The Routing Problem in Traffic Control Using Genetic Algorithms. IFAC Proceedings Volumes. 1994;27(3):187-91.

[8] Mohammed MA, Abd Ghani MK, Hamed RI, Mostafa SA, Ahmad MS, Ibrahim DA. Solving vehicle routing problem by using improved genetic algorithm for optimal solution. Journal of Computational Science. 2017;21:255-62.

[9] de Oliveira da Costa PR, Mauceri S, Carroll P, Pallonetto F. A Genetic Algorithm for a Green Vehicle Routing Problem. Electronic Notes in Discrete Mathematics. 2018;64:65-74.

[10] Liu R, Jiang Z. A hybrid large-neighborhood search algorithm for the cumulative capacitated vehicle routing problem with time-window constraints. Applied Soft Computing. 2019;80:18-30.

IOP Conf. Series: Materials Science and Engineering 1071 (2021) 012025 doi:10.1088/1757-899X/1071/1/012025

[11] Sethanan K, Jamrus T. Hybrid differential evolution algorithm and genetic operator for multi-trip vehicle routing problem with backhauls and heterogeneous fleet in the beverage logistics industry. Computers & Industrial Engineering. 2020;146:106571.
[12] Ibrahim MF, Masudin I, Saputro T. A Hybrid Genetic Algorithm Implementation For Vehicle Routing Problem With Time Windows. Jurnal Ilmiah Teknik Industri. 2015;14:196-204.
[13] Ho W, Ho GTS, Ji P, Lau HCW. A hybrid genetic algorithm for the multi-depot vehicle routing problem. Engineering Applications of Artificial Intelligence. 2008;21(4):548-57.

A Survey Blockchain and Smart Contract Technology in Government Agencies

E S Negara[1*], A N Hidyanto[2], R Andryani[1], D Erlansyah[1]
[1]Faculty of Computer Science, Universitas Bina Darma
[2]Faculty of Computer Science, Universitas Indonesia

Email: e.s.negara@binadarma.ac.id

Abstract. The development of Blockchain and Smart Contract technology provides significant changes in the form of solutions to social technology problems, such as the problem of document forgery on e-government services. In its implementation of government services, there are several challenges faced, such as social technology problems, blockchain mechanisms, virtual machines, and smart contract code levels, and automation of disintermediation. This paper summarizes and discusses the methodology and research results from various recent papers on the state of the use of smart contracts in e-government services and the weaknesses and potential problems in the use of e-government services.

1. Introduction

The contribution and implication of implementing blockchain technology is the improvement of data transaction and communication methods in database systems, where the transaction and data communication methods on the blockchain are decentralized and distributed, where each block contains a cryptographic hash to form a network. The advantages of blockchain technology that can deal with problems such as codification, security, privacy, performance, and traceability are the variables that cause the adoption of this technology to become increasingly popular.

The development of blockchain technology has contributed a lot to various sectors such as: transport and logistics, technology, media, telecommunications, and public sector services, including e-government services, see Figure 1[1].

One of the trends in the development of decentralized and distributed database system technology is the development of blockchain technology which has facilitated the birth of smart contracts[1]. Smart contracts are computerized transaction protocols that independently carry out contract terms and aim to bind agreements or agreements between multiple parties[1][2]. The main characteristics of smart contracts are disintermediation and transparency. Smart contracts are claimed to increase the efficiency of commercialization of a product, reduce transaction costs, eliminate legal problems, and avoid anonymous transactions [1].

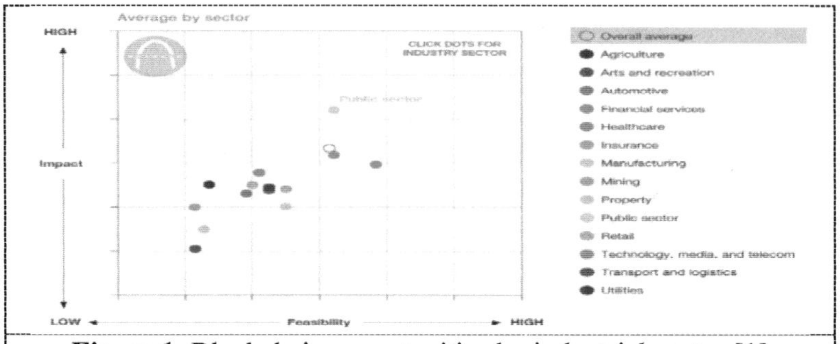

Figure 1. Blockchain opportunities by industrial sector [1]

Various business organizations continue to actively conduct research on the use of blockchain technology for commercial purposes [1]. Likewise government agencies. Currently, various blockchain technology development projects are continuously being developed. At least more than 100 blockchain technology development projects were carried out in the development of government systems in 30 countries [3]. In Indonesia, this technology is also one of the priorities of national research with the theme of developing systems/platforms based on open source with the topic of information technology systems, e-Government communications, artificial intelligence and data analytics [22].

The rapid development of blockchain technology is due to the fact that this technology is directly related to social organizations. Unlike other technologies with traditional consensus. Blockchain works by a consensus algorithm that is pre-agreed upon and after that cannot be modified or faked [3]. Behind the benefits, there are several challenges related to the issue of technology adoption, security and reliability of blockchain technology. In addition, the adoption of smart contract technology also has challenges in adapting the legal framework in a country. Therefore the main research question is:

RQ 1. What is the state of smart contract utilization in e-government service?

RQ 2. What are the Weaknesses and Potential Problems in utilization in e-government service?

This article will discuss the state of the art blockchain technology and smart contract, reviewing the benefits of using them and the potential problems that can arise from using smart contracts in government agencies.

2. Literature Review

2.1 Blockchain Technology and Smart Contracts

The basic concept of blockchain is a distributed database of public records where all digital transactions are executed and shared with participating parties. Each transaction is verified by the consensus of the majority of participants in the system. Records entered in the blockchain system can never be deleted. One of the popular blockchain products is Bitcoin. Bitcoin is a decentralized peer-to-peer digital currency. In its development, digital currencies such as Bitcoin have met with controversy in both financial and non-financial applications. But blockchain has a strong basis so that the controversy can be controlled [15]. Blockchain has five main characteristics, namely: a decentralized database, peer-to-peer transmission, transparency of encryption, permanent data recording, and digital programming based.

In another article, blockchain is a distributed digital transaction and data management system where all users of the system have one consensus or agreement together [15][16][17]. With a distributed system, blockchain eliminates the role of intermediaries which impact the cost of transacting data which is cheaper and safer from data/information manipulation attacks [18]. The mechanisms that make blockchain safe [19] are: Hashing with fingerprint codes; Proof of work; Distributed P2P network. See Figure 1 which is a Blockchain Chain.

IOP Conf. Series: Materials Science and Engineering 1071 (2021) 012026 doi:10.1088/1757-899X/1071/1/012026

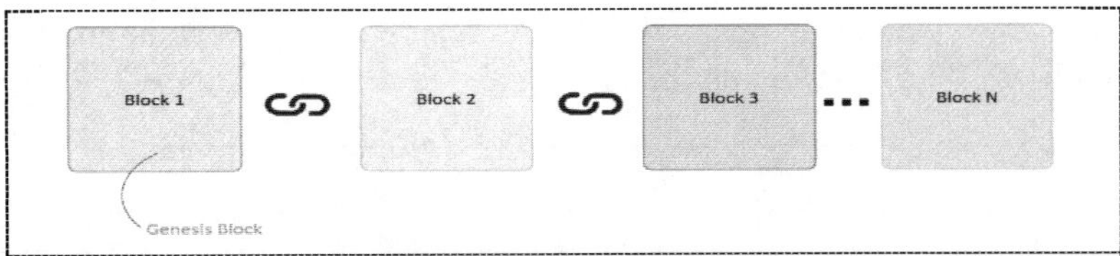

Figure 2. Blockchain chain

Myungsan Jun explained blockchain features with the concept of "social technology" [3]. Social technology is interpreted in a way to communicate, cooperate, compromise, and make agreements between several parties [3]. In addition, social technology also consists of the division of labor, social institutions, and decision-making processes. This technology directly affects the structure of society, systems, social relations, and individual interactions [4].

In its application, physical technology and social technology are related. Where physical technology can praise social technology and allow for the creation of new social technologies. For example, internet technology eliminates geographic boundaries in communicating. Therefore, various studies continue to be conducted to develop social technology based on physical technology interventions. Satoshi Nakamoto's research on blockchain technology is an example of this case [5] [6].

In its development, blockchain technology is expected to change social organizations including the bureaucracy, organizational forms, and the domain of modern society. This is because blockchain is an individual mediation technology and is considered a coercive technique that limits individual activity. Individual mediation is defined as the interaction with each other determined by technology [5] [6].

Bolckhain technology is continuously developed with the concept of automatic algorithm society with artificial intelligence and big data. This technology application is software that can provide logic and control, especially in limiting how individuals behave and interact with this technology. Therefore, Professor Lawrence Lessig formulated the proposition "Code is Law" or "Code as Law" to describe this idea [7].

One of the products of blockchain technology is smart contracts. In particular, the features of a smart contract are considered to have great potential for development. Because smart contracts can set rules and define conditions for contract execution automatically. The advantages of blockchain technology implemented in smart contracts can prevent hacking or contract forgery. So that with this concept it will be possible to build an absolute law [8].

Smart contracts are digital coded containers that encode real-world contract deals. The main element of a contract is an agreement that binds two or more parties. Where each party has obligations based on the agreement. Another element is that the agreement is mediated by a centralized legal entity or can be referred to as a third party. However, with a smart contract, third party functions can be eliminated because both parties already have a digital contract key. Digital contract lock is an automatic code execution that is distributed and verified by a decentralized blockchain network system. Therefore smart contracts are considered as the latest technology that can be trusted to eliminate dependence on third parties and eliminate third party commission fees [9].

Another definition of a smart contract is an autonomous agent. Smart contracts as autonomous agents are stored on the blockchain which are recorded by coding creation transactions then contracts are introduced to the blockchain. Then the smart contract is identified by the contract address. Each smart contract has a sum of virtual coins (Ether), has its own store, and is associated with a predefined executable code. The contract status consists of two main parts: personal storage and the amount of virtual coins (Ether) which is referred to as balance [20].

Smart contracts are developed on top of the cryptocurrency platform. Cryptocurrency is a decentralized cryptosystem that deals with a shared global ledger in the form of virtual money. Users can transfer money and interact with contracts according to agreements signed via cryptocurrency network transactions. The network consists of mining nodes in charge of disseminating information, storing data, and updating data. The smart contract scheme is shown in Figure 2 [21].

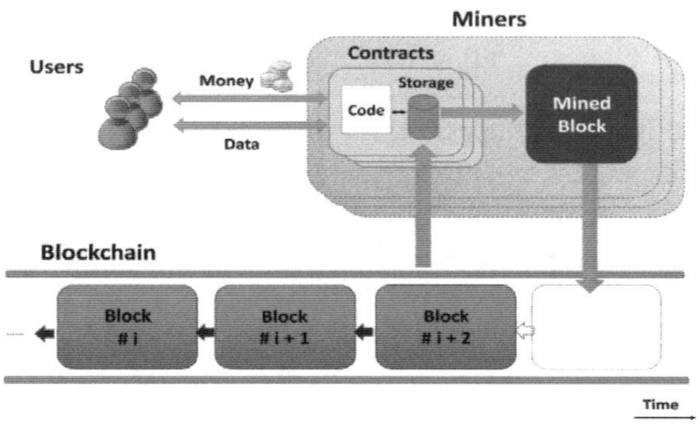

Figure 3. Cryptocurrency Decentralized Scheme on Smart Contracts [21].

Smart contracts consist of the following platforms:

- Ethereum. The biggest and most popular platform for smart contracts is Ethereum. Ethereum smart contracts are utilized in many digital transactions such as building distributed applications, building social networks and identity systems, predicting markets, and various types of financial applications.
- Bitcoin. Bitcoin smart contract technology is a cryptocurrency system that provides new ways to increase the circulation of capital, new markets, and new decentralized autonomous organizations. Bitcoin uses the OP_RETURN transaction of the bitcoin scripting language to store metadata on the blockchain.

Apart from Bitcoin and Ethereum, some new platforms have emerged from either the original Bitcoin blockchain or standalone and offer new enhancements and approaches to the bottlenecks found in the previous platform.

3. Discussion

3.1 Utilization of Blockchain Technology and Smart Contracts in E-Government Services.

Blockchain technology and smart contracts have been widely used in various projects carried out by governments in various countries. There are at least 17 countries that are undergoing various projects by the government (apart from voting systems and digital currency projects) such as; in the United States (US) the government is developing an online personal health data exchange, in the UK through the Department of Labor and Pensions is testing a welfare payment system that carries the concept of blockchain, and in China is developing an asset storage system with the main base of blockchain technology with capacity of 100 transactions in October 2016, as well as various other projects in several countries that the author believes are still in the planning stages or progress. In addition, there are other uses in government agencies in the form of blockchain-based electronic voting system projects, and blockchain-based digital currency projects around the world [3].

The systematic literature review conducted by Yli-huumo et al summarizes the general concept of blockchain technology that has been developed by various researchers [10]. The study shows that 80% of bitcoin development projects focus on security and privacy themes. And since 2016 blockhain applications have diversified. Therefore Yli-huumo et al conducted a research that focuses on cybersecurity and blockchain applications [10].

Meanwhile, at the end of 2016 Conoscenti et al also conducted a systematic literature review on the use and adaptability of blockchain, especially to the Internet of Things (IoT) and several other peer-to-peer devices [11]. The results of these studies show that blockchain can be used for data misuse detection without the need for a central reporting mechanism [11]. A literature review was also carried out by Seebacher and Schüritz [12]. The results of his study show the blockchain service system plays an integral role in facilitating shared value creation, ensuring the availability of information and providing a system coordination mechanism [12].

According to research conducted by Myungsan Jun [3], what makes many countries start blockchain projects is because it is different from other traditional technologies where the consensus mechanism is not a machine but a human domain, blockchain technology works with a human consensus algorithm that functions to decide what data is considered genuine. then stored on the blockchain. This technology allows all parties to validate the data and save the original verified version. Verified data cannot be changed or falsified.

The other main benefits of implementing blockchain technology in government are claimed to be as follows [13]:

- Reducing the economic costs, time, and complexity of exchanging information between government and public-private which improves the functioning of government administration.
- Eliminate bureaucratic channels, eliminate discretionary power, and prevent corruption. This is possible because of the use of distributed nested books and short-cut contracts that have been programmed before.
- Can improve the automation, transparency, auditability and accountability of information in government services.
- Can increase trust in organizational governance through the use of smart contract algorithms which are under government control

Thus, blockchain is a state-of-the-art social and physical technology that simultaneously enables a system that is immutable and tamper-resistant, making it optimal in handling public data that cannot be forged. With the smart contract feature, this technology has the potential to replace existing social organizations, making blockchain one of the most innovative digital technologies that can be considered in the new paradigm of government policymaking and service delivery.

3.2 Weaknesses and Potential Problems

After discussing the state of the art regarding the concept of blockchain technology, printar contracts and the use of this technology in government services, this article also discusses the weaknesses and potential problems of smart contract law in the case of technology adoption and adaptation. As is well known, almost all technologies are vulnerable to cyber attack threats. Cybercrime has caused many economic losses, especially losses that occur due to cyber attacks on digital commercial transactions, theft of personal information and digital financial threats. This condition causes the risk of security breaches to continue to increase exponentially [1].

Utilization of smart contracts requires the involvement of each party in the digital transaction contract being carried out. This can threaten the risk of information transaction sensitivity. In addition, hackers have a lot of skills and adaptability to the system today. So that it allows opportunities for threats to system attacks such as blockchain [1]. The increasing intention of cyber crime is partly due to the increasing value of virtual currencies and the increasing use of smart contracts. This condition inspires cyber criminals to steal digital money, ask for ransom and carry out illegal transactions [1].

The results of a systematic mapping study conducted by Macrinici D, et al, which summarizes several studies on smart contracts on blockchain technology that show at least 16 smart contract problems. All of these issues are categorized into three main issues: blockchain mechanisms, virtual machines, and smart contract code level. From the results of the study, the biggest problem is the problem of blockchain mechanism and the source code of smart contracts. This is due to the complexity of the blockchain, which consists of scalability, security, privacy, and human error invoices. Human error factors include lack of knowledge, programming errors, and typing errors [14].

Another problem that can arise from this technology is related to the main rationale of smart contracts, namely disintermediation automation. This concept is a contract between parties that execute themselves without an intermediary trusted to facilitate transactions or agreements such as non-digital commercial transactions (traditional functions) which are usually mediated by legal persons or financial authorities. The traditional function of financial experts and commercial lawyers today can also be carried out using smart contracts so that they can replace their roles that are usually profitable to become disadvantaged. Some argue that the financial and legal workforce may suffer losses due to unreliable blockchain technology [14]. However, the authors argue that smart contracts will not pose an immediate and serious threat as feared. There will still be room in various sectors for professionals and lawyers who remain deeply embedded in our global economy because they can think in ways that computers cannot.

4. Conclusion

Blockchain is a cutting-edge social and physical techniques that can solve problems that often occur in handling public data such as data falsification. By utilizing the smart contract feature in e-government, this technology can solve existing social problems, so blockchain and smart contracts are among the most innovative digital technologies that can be considered in the new paradigm of government policymaking and service delivery. In implementing and developing smart contract technology for e-government services, apart from social technology problems, there are at least 4 main challenges that may occur, namely the blockchain mechanism, virtual machines, and smart contract code level, and disintermediation automation.

Acknowledgments

Thank you to the Ministry of Research and Technology/National Research and Innovation Agency for research funding assistance through the Post-Doctoral Research scheme in 2020.

References

[1] Mark Giancaspro, Is a 'smart contract' really a smart idea? Insights from a legal perspective, Computer Law & Security Review: The International Journal of Technology Law and Practice (2017), doi: 10.1016/j.clsr.2017.05.007. Hal 1-10

[2] Watanabe, H., Fujimura, S., Nakadaira, A., Miyazaki, Y., Akutsu, A., & Kishigami, J. (2016, January). Blockchain contract: Securing a blockchain applied to smart contracts. In 2016 IEEE international conference on consumer electronics (ICCE) (pp. 467-468). IEEE.

[3] Myungsan Jun. Blockchain government - a next form of infrastructure for the twenty-first century. Journal of Open Innovation: Technology, Market, and Complexity. (2018) 4:7. DOI 10.1186/s40852-018-0086-3. Hal 1-12.

[4] Nelson, R.R. & Nelson, K. (2002) "Technology, institutions, and innovation systems", Research Policy.

[5] Dougherty, D. (2017) 'Taking advantage of emergence for complex innovation eco-systems', "Journal of Open Innovation: Technology, Market, and Complexity", Keynote paper for SOItmC & Riga Technical University 2017 Conference. https://doi.org/10.1186/s40852-017-0067-y.

[6] Nakamoto, S. (2008) "Bitcoin: A Peer-to-Peer Electronic Cash System", https://bitcoin.org/bitcoin.pdf

[7] Lessig, L. (2000) "Code: And Other Laws of Cyberspace", Basic Books.

[8] Tapscott, D.,·Tapscott, A. (2016) "Blockchain Revolution", Penguin Random House UK.

[9] Swan, M. (2015) "Blockchain: Blueprint for a New Economy", O'Reilly Media.

[10] J. Yli-Huumo, D. Ko, S. Choi, S. Park, K. Smolander, Where is current research on Blockchain technology? - a systematic review, PLoS One 11 (10) (2016) 127.

[11] M. Conoscenti, A. Vetr, J.C. De Martin, Blockchain for the internet of things: a systematic literature review, in: 2016 IEEE/ACS 13th International Conference of Computer Systems and Applications (AICCSA), 2016, p. 16.

[12] S. Seebacher, R. Schritz, Blockchain technology as an enabler of service systems: a structured literature review, in: Exploring Services Science, 2017, p. 1223.

[13] David Allessie, Maciej Sobolewski, Lorenzino Vaccari. Blockchain for digital government, An assessment of pioneering implementations in public services. Publications Office of the European Union, Luxembourg, 2019, ISBN 978-92-76-00581-0, doi:10.2760/942739, JRC115049

[14] Macrinici, D., Cartofeanu, C., Gao, S., Smart contract Applications within Blockchain Technology: A Systematic Mapping Study, Telematics, and Informatics (2018), doi: https://doi.org/10.1016/j.tele.2018.10.004.

[15] Crosby, M., Pattanayak, P., Verma, S., & Kalyanaraman, V. (2016). Blockchain technology: Beyond bitcoin. Applied Innovation, 2(6-10), 71.

[16] Iansiti, M., & Lakhani, K. R. (2017). The truth about blockchain. Harvard Business Review, 95(1), 118-127.

[17] Yli-Huumo, J., Ko, D., Choi, S., Park, S., & Smolander, K. (2016). Where is current research on blockchain technology?—a systematic review. PloS one, 11(10), e0163477.

[18] Xu, J. J. (2016). Are blockchains immune to all malicious attacks?. Financial Innovation, 2(1), 25.

[19] Husni(2019). Mengenal blockchain: Teknologi dibelakang Bitcoin. Komputasi. 3(1).

[20] Luu, L., Chu, D. H., Olickel, H., Saxena, P., & Hobor, A. (2016, October). Making smart contracts smarter. In Proceedings of the 2016 ACM SIGSAC conference on computer and communications security (pp. 254-269). ACM.

[21] Delmolino, K., Arnett, M., Kosba, A., Miller, A., & Shi, E. (2016, February). Step by step towards creating a safe smart contract: Lessons and insights from a cryptocurrency lab. International Conference on Financial Cryptography and Data Security (pp. 79-94). Springer, Berlin, Heidelberg.

[22] Negara, ES., Kerami, D., Wiryana, IM., & Maulana Kusuma TB (2017). Researchgate data analysis to measure the strength of Indonesian research. Far East Journal of Electronics and Communications, 2017, 17(5), pp. 1177–1183.

IOP Conf. Series: Materials Science and Engineering 1071 (2021) 012027 doi:10.1088/1757-899X/1071/1/012027

Multiple Smile Detection Using Histogram of Oriented Gradient and Support Vector Machine Methods

R Syaputra[1*] D Syamsuar[1] and E S Negara[1]
[1]Computer Science, Universitas Bina Darma Palembang, Indonesia

E-mail: rezkisyaputraharahap@gmail.com

Abstract. The face or the front of the head consists of the eyes, nose and mouth. Each face has its uniqueness from the many human faces. The face is used to show happy and sad expressions and feelings. Smiling also includes self-expression from others. The analysis of facial expressions plays a key role in analyzing human emotions and behavior. Smile detection is a specialized task in facial expression analysis with a variety of potential applications such as photo selection, user experience analysis, smile payments, and patient monitoring. Conventional approaches often extract low-level face descriptors and smile detection based on strong binary classifiers. In this paper, we propose an effective Histogram of Oriented method supporting vector engines for smile detection. The experimental results show that our proposed network outperforms the current state of the method. The test results using the Histogram of Oriented Gradient and Support Vector Machine method for smile detection were 87% for a precision value of 88% and a recall value of 83% and accuracy. In the future, we want to exploit some of the latest effective designs. we will try to update our mouth model so we can support a bigger head turn and face size scale.

1. Introduction

Common facial expressions in everyday life are smiles that express emotional feelings and underlie feelings such as joy, happiness, sadness, boredom, anger, satisfaction, and dissatisfaction. The need for smile monitoring is of importance and challenges the development and advancement in the real world to assist humans in various potential applications, such as photo selection, user analysis, payment using a smile, applications in the healthcare industry, application of artificial intelligence, and patient feeling conditions. Smile detection has already attracted the interest and attention of researchers, professionals in the development of this new science.

The efficient approach and development method used is the difference in data intensity between pixels in the image database using a simple grayscale feature. Then a figure database was adopted to select and combine the devices using a classification method based on pixels to form a robust classification for smile detection.

This study, representing a face using three artificial features, 1) Local Binary Pattern (LBP), 2) Local Phase Quantization (LPQ), 3) Histogram of Oriented Gradient (HOG). Use the gradient orientation histogram method to represent image data (faces) and to classify image data (faces) using the support vector machine method.

In recent years, researchers have witnessed many great technological achievements in computer science. The number of smart technologies that have been successfully created by researchers and professionals and combine computer science and other sciences that provide concrete examples such as IBM Watson, Health robots, Google AI driverless cars, and delivery systems using drones on Amazon developed in advances in computer technology. In the advancement of computer technology, it is necessary to have in-depth learning about computer technology that can help humans, especially using the Convolutional Neural Network (CNN) method, this method contributes to the development of this technology. Convolutional neural networks are considered to be the deepest and most effective neural network architectures for data classification such as images. Researchers have started implementing a convolutional neural network to detect smiles. a convolutional neural network that exists to extract higher-level features from an image database (faces). Uses a classification support vector machine to detect smiles for classification tasks. This study aims to develop and apply artificial intelligence-based methods to recognize the object of a smile. The development of this research is very useful for detecting the object of a smile that can be used in various ways.

2. Literature Review

2.1. Smile

Smile is a facial expression to show someone's feeling's whether someone is sad, happy and so forth. There are two types of smile, one social smile is a voluntary facial expression that is intentionally made like when getting acquainted with others, making a passport photo, two spontaneous smiles are facial expressions that are accidentally made, natural and caused by emotional overflow. Smiles in each person are different where the shape of each person's smile is not same but the smile has the same pattern, where the pattern of the right and left edge lip is higer than the middle lib. There are some extraction features for detecting smiles such as haar-like feature extraction over gray colors and standardized positions such as hear done by pinky and manihs, histogram equalization refers more to the color of each block done by eduard and ivanna, features gabor extraction is more directed toward frequency and the edge detections extraction feature which is more directed towards the outline of a shape is done by eduard in the same year.

The use of haar-like features, histogram equalization and edge detection is susceptible to light interference even though it has a pretty good shape recognition result, which is for haar-like ti get 85% accuracy which is better than histogram equalization which has an accuracy of 50% and gabor 81,6%, while the use frequencies allows misleading to identify features. While the hog feature has a better shape recognition ability which is shown by aryuanto therefore this study uses hog ectraction feature which aims to recognize the shape of a smile.

2.2. Support Vector Machine

Support Vector Machine is looking for the best hyperplane that functions as a separator of two data classes to form a model that will be used in the classification process or Support Vector Machine (SVM) is a binary classification method that utilizes a hyperplane equation that divides the two classes like (Figure 2). The two label classes or y in SVM are $\in \{-1,1\}$. Data xd is a feature vector with d = 1,2, ..., Nd dimensions. For $d = 2$ the hyperplane is a line, while for $d = 3$ it is a plane. Hyperplane equation is a classification equation shown by equation [12] with the classification parameter is the value w is the weight value and b is biased.

$$f_{svm}(x) = w.x + b$$

To find the values of w and b, equations (13) and (14) are used

$$b = -\frac{1}{2} w.x^{+} + w.x^{-}$$

$$w = \sum_{i=1}^{N} a_i \, y_i \, x_i$$

N: the amount of data
α: coefficient that functions or is used to find weight values
y: a label for each lip
x: is the feature vector of the HOG.

Support Vector Machine is also used to manage datasets by utilizing kernel tricks to find hyperplanes, one of which is the Linear Kernel shown by equation in [13]-[15].

$$K(x_i, x_j) = x_i, x_j$$

Place the Linear Kernel Support Vector Machine Visualization image From the kernel results obtained, Support Vector Machine makes a classification equation that is adjusted to the kernel used, such as equation (16) with class classification values based on equation [16]-[17].

$$f_{svm}(x) = \sum \alpha_i y_i K(x_i, x) + b$$

$$class = \begin{cases} 1, f_{svm}(X) \geq 0 \\ -1, selanjutnya \end{cases}$$

The determination of the value of α_i is done by solving the LD (Lagrange Multiplier Duality) optimization problem, which is shown by equation (18) and by the formula in the equation.

$$Ld = \frac{max}{\alpha i \in R} \sum_{i \in N} \alpha_i \sum_{i \in N} \alpha_i \alpha_j y_i y_j$$

Which has two conditions 0 ≤ α_i ≤C and ∑_ (i∈N) $\sum_{i \in N} \alpha_i y_i = 0$where C is the constant value with, Ld (x_i), the kernel function used.

2.3. Histogram of Oriented Gradient (HOG)

Histogram of Oriented Gradient (HOG) is the extraction of features used in computer vision and image processing by calculating the gradient value in an image to get the results that will be used to detect the objects. Histogram of Oriented Gradient is a form of local object and a value that is used from the gradient intensity. Histogram of Oriented Gradient (HOG) is the extraction of features used in image processing computers by calculating the gradient value in an image to get the results that will be used to recognize the characteristics of the object.

The histogram of Oriented Gradient is the shape of the object's locale and the value used for the intensity of the gradient. The process in using Histogram of Oriented is to look for the gradient orientation and vertical gradient values and then look for the magnitude and binning orientation values of the original image size then divide the image into several blocks that have a size of 2x2 which later in the block there are several cells with a size of 8x8 which have a gradient orientation 9 bin so that the feature vector is obtained, each cell block has a direction gradient value or an edge orientation value for each image pixel. In improving the performance of the resulting gradient values carry out a binning orientation by normalizing in contrast, then this value is used to describe each block of the results of the normalized value. The first stage of the Histogram of Oriented process is to find the value of the gradient calculation using equation [1][2].

$$g_x(r,c) = g(r,c+1) - g(r,c-1)$$
$$g_y(r,c) = g(r+1,c) - g(r-1,c)$$

Or use the Central Difference Gradient Operator equation:

$$g_x = [-1\ 0\ 1]$$

$$g_y \begin{bmatrix} 1 \\ 0 \\ 1 \end{bmatrix}$$

Where g = gradient r = row c = column

After the x and y values have been obtained, the next process is to find the magnitude μ value to find out how big the gradient is and look for the value bin binning orientation using equations [3] and [4].

$$\mu = \sqrt{g_x^2 + g_y^2}$$

$$\theta = \frac{180}{\pi}\ (tan_2 -^1 (ly, lx)\ mod\ \pi$$

After getting the θ value of each cell in the block, then look for the histogram for each cell for bin j, Cid a where the value of B is determined from the number of bins, namely [9] bin using equations [5] and [6].

$$bin\ j = \left[\frac{\theta}{w} - \frac{1}{2}\right] mod\ B$$

$$w = \frac{180}{B}$$

Then the value obtained does vote bin which is divided into two types namely vote bin Vj indicated by equation [7] and vote bin (+ 1) indicated by equation [8] and the boundary of the bin is shown in equation.

$$Vj = \mu \frac{c_j +_1 -^\theta}{w}$$

$$V_{j+1} = \mu \frac{\theta - c_j}{w}$$

With Wi-Fi Limits, W (i + 1)

$$C_i = W\left(i + \frac{1}{2}\right)$$

Later, the temple is looking for block distribution. If a vote has been obtained from each bin, then normalize the block which is used to get the b value of each block to reduce the change in contrast between blocks of objects to find block normalization using equation [10].

$$b \leftarrow \frac{-b}{\sqrt{||b||2_{-\varepsilon}}}$$

After getting the block value, the next step is to look for the Histogram of Oriented feature where the normalized blocks are combined into one Histogram of Oriented feature vector using equation [11].

$$h \leftarrow \frac{-h}{\sqrt{||h||2_{-\varepsilon}}}$$

The process in applying hog is to divide the image into blocks that later in the block there are image. In increasing the performance of the resulting gradient value, the binning orientation is done by normalizing in contrast, then this value is used to describe each block from the normalized value. Following the implematation of Histogram of Oriented can be seen in Fig 1.

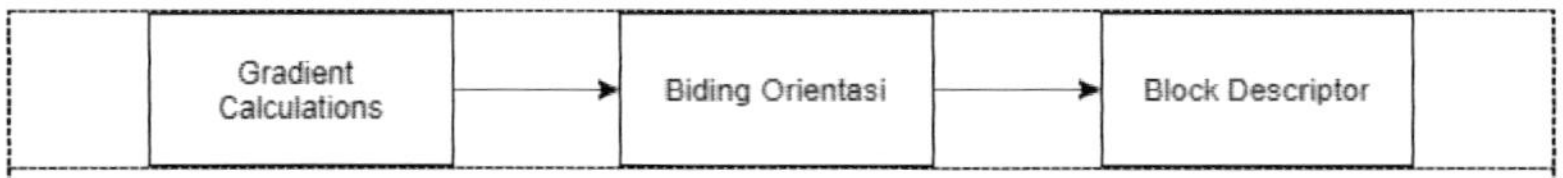

Figure 1. . Implementation of Histogram of Oriented

In fig 1, the steps of Histogram of Oriented implementation are starting from gradient calculation, binning orientation, then block descriptor by Dalal, Trig and Derry, while the next stage in the Histogram of Oriented process is finding the gradient calculation value first to get value of gradient calculation using equation [1][2].

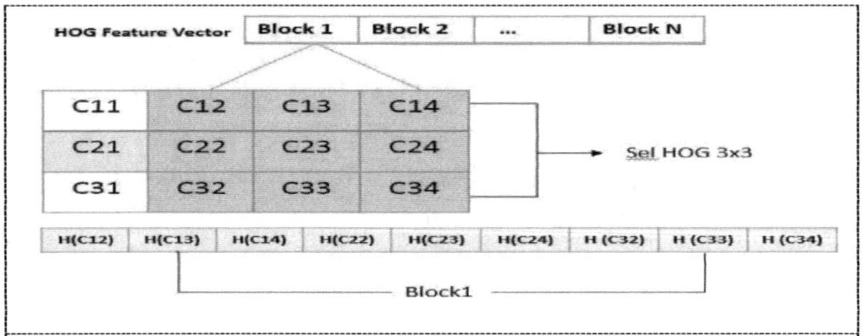

Figure 2. Division of Block for Each Cell

Fig 2 is a block division scheme for each cell, where this value will be used as a block val the probability of H (prior), P (X | H) is the probability of X in the hypothesis H, P (X) is the probability of X.

3. Results and Discussion

3.1. Framework

At the initial stage, the face image performs a grayscale process to speed up the duration of the program process and then continues at the image extraction stage Histogram of Oriented uses the existing functions in python and the results of the image extraction enter the Support Vector Machine training stage, this model is to load Support Vector Machine structs and groups. After the Support Vector Machine model training process is complete, the Support Vector Machine model data has been obtained and can be used to test data. Here are some steps are taken to identify multiple smiles using the Histogram of Oriented (HOG) Method and Support Vector Machine (SVM):

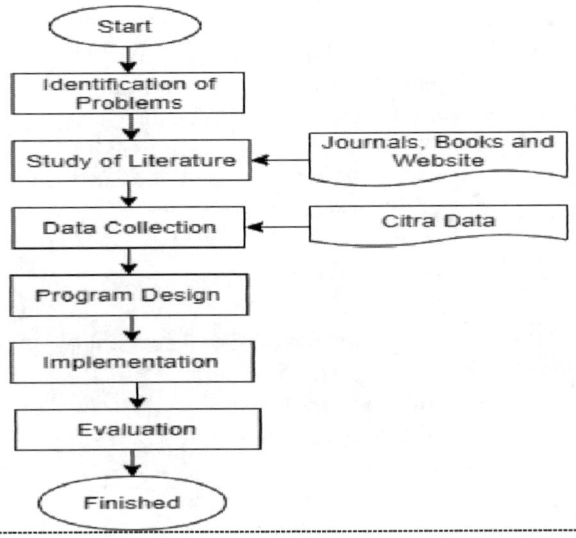

Figure 3. Framework

Given a set of pictures of a person into the system, it compares these images and determines which one has the best smile. Images can be from the same person or from several different individuals. To identify the subject's face in the photo using and the same algorithm is used to locate the subject's mouth. Furthermore, the Shi-Tomasi angle detection algorithm is carried out through the mouth area; it locates the edges and features of the mouth (smile wrinkles, teeth, mouth shape). It uses the points obtained from the detection angle; the most suitable second-degree polynomial line is drawn. By taking the derivative of the most suitable line it determines the concave point, and from there one can detect whether the subject is smiling or not.

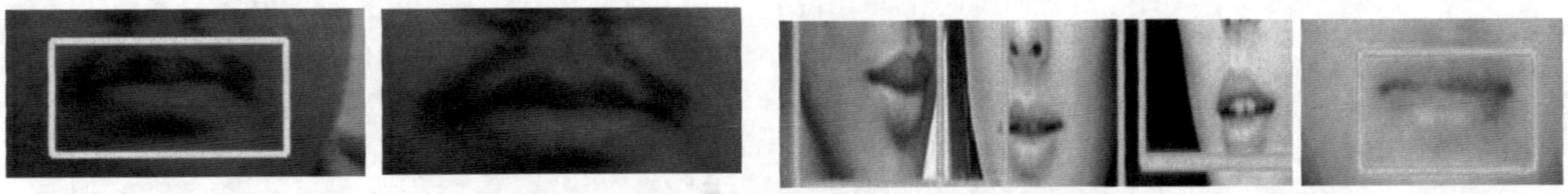

Figure 4. Smile Detection results

Our methods and algorithms are implemented using Python. The researcher tested the algorithm using facial images from the database. The database contains color images of the front faces with different lighting, expressions and backgrounds. From the testing process, the researcher found that our proposed method shows the best detection rate, both lip detection rate and smile detection rate. This is because some of the lip areas obtained with the mouth detector are too small (they do not cover allof the mouth/lips). So that a smile cannot be detected properly.

4. Conclusion

This study uses a gradient-oriented histogram method and a supporting vector machine to assist in handling smiles. Previous research used convolutional neural networks. Several previous studies used shallow convolutional neural networks with few hidden layers. Apart from that, this research also shows inefficiency and data argumentation techniques (figure) in overcoming lack of data to avoid overfitting.

By detecting the characteristics and the angle of detection, we can detect whether the person is smiling or not in a photo with Face Detection. This Automatic Smile Identification has many potential applications, including enhanced camera functionality for advanced functionality. The test results using the Histogram of Oriented Gradient and Support Vector Machine method for smile detection were 87% for a precision value of 88% and a recall value of 83% and accuracy. In the future researchers want to develop some of the model more effective. we will try to update our mouth model so we can support a bigger head turn and face size scale.

References

[1] L. An, S. Yang, and B. Bhanu. Effecient smile detection by extreme learning machine. *Neurocomputingi*, 149:354-363, 2015.

[2] V. Jain and J. L. Crowley. Smile detection using multiscale gaussian derivatives. In 12th WSEAS *International Conference on Signal Processing, Robotics and Automation*, 2013.

[3] C. Junkai, O. Qihao, C. Zheru, and F. Hong. Smile detection in the wild with deep convolutional neural network. *Machine Vision and Applications*, 28(1):173–183, 2017.

[4] M. Liu, S. Li, S. Shan, and X. Chen. Enhancing recognition in the wild with unlabeled referencedata. *In Asian Conference on Computer Vision*, pages 577–588. Springer, 2012.

[5] Y. Mardi, "Data Mining : Klasifikasi Menggunakan Algoritma C4.5," *J. Edik Inform.*, vol. 2, no. 2, p. 7, 2017.

[6] C. Shan. Smile detection by boosting pixel differences. *IEEE transactions on image processing*, 21(1):431–436, 2012

[7] Imagenet large scale visual recognition challenge 2014 (ilsvrc2014).*http://www.imagenet.org/challenges / LSVRC/2014/.* Accessed: 2017-06-15.

[8] M. B. Pranoto dan K. N. Ramadhani, "Face Detection System Menggunakan Metode Histogram of Oriented Gradients (HOG) dan Support Vector Machine (SVM) Face Dtection System using Histogram of Oriented Gradients (HOG) Method amd Support Vector Machine (SVM)," *e-Proceeding Eng.*, vol. 4, no. 3, hal. 5038–5045, 2017.

[9] D. Amputri, S. Nadra, G. Gasim, dan M. E. Al Rivan, "Perbandingan jarak potret dan resolusi kamera pada tingkat akurasi pengenalan angka kwh metermenggunakan svm," *J. Inform. Glob.*, vol. 8, no. 1, hal. 7–12, 2017.

[10] L. Cunhe and W. Chenggang, "A new semi- supervised support vector machine learning algorithm based on active learning," in *2010 2nd International Conference on Future Computer and Communication*, Wuhan, China, 2010, pp. V3-638-V3-641, doi: 10.1109/ICFCC.2010.5497471.

[11] M. Hofmann, "Support Vector Machines — Kernels and the Kernel Trick," *Notes*, vol. 26, no. 3, p. 16, Jun. 2006.

[12] C. Manning, P. Raghavan, and H. Schuetze, "*Introduction to Information Retrieval,*" p. 581, Apr. 2009.

[13] T. Sutabri, A. Suryatno, D. Setiadi, and E. S. Negara, "Improving Naïve Bayes in Sentiment Analysis For Hotel Industry in Indonesia," in *2018 Third International Conference on Informatics and Computing (ICIC)*, Palembang, Indonesia, Oct. 2018, pp. 1–6, doi: 10.1109/IAC.2018.8780444.

[14] M. Raza, F. K. Hussain, O. K. Hussain, M. Zhao, and Z. ur Rehman, "A comparative analysis of machine learning models for quality pillar assessment of SaaS services by multi-class text classification of users' reviews," *Future Gener. Comput. Syst.*, vol. 101,

pp. 341–371 Dec.2019, doi: 10.1016/j.future.2019.06.022.

[15] C. Darujati and A. B. Gumelar, "Pemanfaatan Teknik Supervised Untuk Klasifikasi Teks Bahasa Indonesia," *J. Link*, vol. 16, no. 1, p. 8, Feb. 2012.

[16] K. Zhang, Y. Huang, H. Wu, and L. Wang. Facial smile detection based on deep learning features. *In Pattern Recognition (ACPR), 2015 3rd IAPR Asian Conference* on, pages 534–538. IEEE, 2015.

[17] C. Szegedy, W. Liu, Y. Jia, P. Sermanet, S. Reed, D. Anguelov, D. Erhan, V. Vanhoucke, and A. Rabinovich. Going deeper with convolutions. *In Proceedings of the IEEE Conference on Computer Vision and Pattern Recognition*, pages 1–9, 2015.

E-money Academic: Lesson from Literature Visualizing Scientometric Positioning (1968-2019)

A Purnomo[1*], M Firdaus[2], U Hasanah[3], A V D Sano[1] and M A Prasnowo[4]
[1]Bina Nusantara University, Indonesia
[2]Universitas Nahdlatul Ulama Sidoarjo, Indonesia
[3]Universitas Trunojoyo Madura, Indonesia
[4]Universitas Ma'arif Hasyim Latif, Indonesia

E-mail: agung.purnomo@binus.ac.id

Abstract. There is no overview of literature in the area of e-money research which shows the big picture using data from all countries This paper aims to review the status and visual map position of research in the e-money literature indexed Scopus that used a scientometric positioning overview. The research was carried out using bibliometric techniques. Data analysis as well as visualization utilizing VOSViewer program and the Scopus function for analyse search results. In this review, the details collected applied to 3,011 documents issued from 1968 through 2019. The study reveal that Miller, A. and Chinese Academy of Sciences were the most active individual scientists and affiliated institutions in e-money literature. In e-money, computer science and Lecture Notes in Computer Science Including Subseries Lecture Notes in Artificial Intelligence and Lecture Notes in Bioinformatics were the most areas of study and dissemination sources. There were nine worldwide group maps with collaborative researchers. In order to identify the body of knowledge created from fifty-one years of literature, this study constructed a convergence axis grouping comprising of e-money: Mobile telecommunication system, Electronic cash system, Finance, Profitability, Investments, Intelligent system, abbreviated as MEFPII themes.

1. Introduction

As a fintech innovation that makes transactions easier for customers, e-money is here. The drastic inoculation of digital technology has an impact on all areas of contemporary life, particularly financial activity, leading to the need to develop new simulation methods and to analyze a large number of links and interactions between various financial agents [1]. Payments via the internet have grown in popularity in recent years. With the support of a website or client application, they are executed [2]. several fintech technologies have been launched that have changed the way of trading and banking, such as digital wallets, payment applications, mobile banking, mobile trading, robo-advisor platforms, and peer-lending sites [3]. As a strategic information system, electronic payment systems, especially in developing countries, are considered one of the key components of economic growth and are very helpful in strengthening the capacity and delivery of financial services [4]. In addition, by providing financial services to the masses without having to have a bank account, digital payments that can be enabled specifically on smartphones can transform the lives of millions of people in developing countries [5].

Electronic money or e-money is a result of technological evolution that provides tools to escape transfers of physical money, known as digital money, mobile money, mobile payments, electronic cash [6]. E-money is characterized as a stored value or prepaid commodity in which someone owns a sum of money in a fund record or value in electronic media [7]. Other definitions of electronic money are: 1) fulfilling property: circulation, no bank accounts, no compensation, universal use; 2) storing electronic devices, issuing funds obtained and received as a means of payment; 3) identifying user demands that are stored and received as payment on the system [8]. E-money is growing with the rise of digital entrepreneurship and technopreneurship.

E-money has benefits, from ease of use to ease of use in purchasing and selling transactions with e-money, including ease of use [9]. The benefits of using e-money include faster and more efficient, safer payment transactions that allow banking services to be better customized [10]. Other benefits of electronic money include: payments made via any cell phone, customers do not need a bank account, do not need the internet, communicate through SMS, and physical money can be exchanged at any time. It is possible to make online orders, service payments, national transfers, collection control, consumer-to-consumer purchases in supermarkets, public utilities [11]. On the other hand, e-money shows little value in a broad economy, namely the key advantages of the Consumer Payment Option Diary are 1) the ability to calculate expenses with payment instruments combined into solid transactions, 2) relatively low expenditures, and 3) relatively low cost-effective expenditures [12].

The use of e-money was initially for practicality, because if you want to buy anything, you don't need to bring cash. As required, e-money cardholders should choose an e-money card. Since the industry has many e-money cards and provides numerous payment facilities. In addition, not all retailers can allow e-money payment transactions or it can be said that no e-money card can fulfill all requirements [13]. There are many benefits of using e-money in the IoT-connected era, however, in general, previous research related to e-money was limited to one country [14], [15]. There is still a need for research to explain this. Unfortunately, despite presenting a broad image map visualized year over year with details from several published studies at the global scale, there has not been much literature on e-money. The strong positive relationship regarding affiliation, scholars, and the impact of scholarly studies has also not been explicitly discussed by any literature. This paper aims to review the status and visual map position of research in the internationally e-money literature indexed Scopus that used a scientometric perspective.

2. Research Methods

This review mapped the status of study conducted in the last 51 years at global level on the basis of e-money. In August 2020, this study collected data from the Scopus utilizing document search queries [16]. The research was carried out using bibliometric techniques. Data analysis as well as visualization utilizing VOSViewer program and the Scopus function for analyze search results [17]–[19]. The VOSViewer tool is used to build and visualize bibliometric networks, including the number of studies, countries, academic affiliations, keywords, scholars, fields and collaborations between writers [20].

This study identifies e-money keywords to recognize and look for Scopus database publications with 3,011 globally published documents from 1968 through 2019. The research confined collection of data to 2019 and excluding 2020. In order to reflect the state of the study over the entire year, the annual academic data collected from January to December. (TITLE-ABS-KEY (e-money) AND PUBYEAR <2020) is the query input command which is implemented while mining academic publication data on online database of Scopus.

The research applies a co-authorship analysis with authors' analysis units and full calculation systematic techniques utilizing VOSViewer to gain the collaboration research network of the international researcher. The research conducted an in-depth co-occurrence analysis with keyword relation analysis as well as a full systematic technique of calculation utilizing VOSViewer to generate a keyword map network.

IOP Conf. Series: Materials Science and Engineering 1071 (2021) 012028 doi:10.1088/1757-899X/1071/1/012028

3. Result and Discussion

E-money literature appear to be likely to increase and grow per year. The tallest point for international publication was 720 documents in 2019. Since 1968, publishing on e-money has already started.

3.1. E-Money Literature Most Common Organizational Affiliations

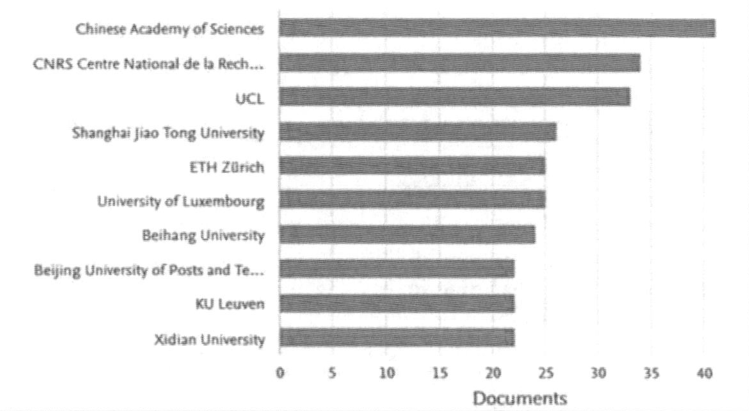

Figure 1. Organizational Affiliation Number of Annual Literature of E-Money

The leading research organizations in e-money literature was Chinese Academy of Sciences with 41 documents. Then followed by CNRS Centre National de la Recherche Scientifique, UCL, Shanghai Jiao Tong University, ETH Zürich, University of Luxembourg, Beihang University, Beijing University of Posts and Telecommunications, KU Leuven, Xidian University and Cornell University.

3.2. E-Money Literature Most Individual Researcher

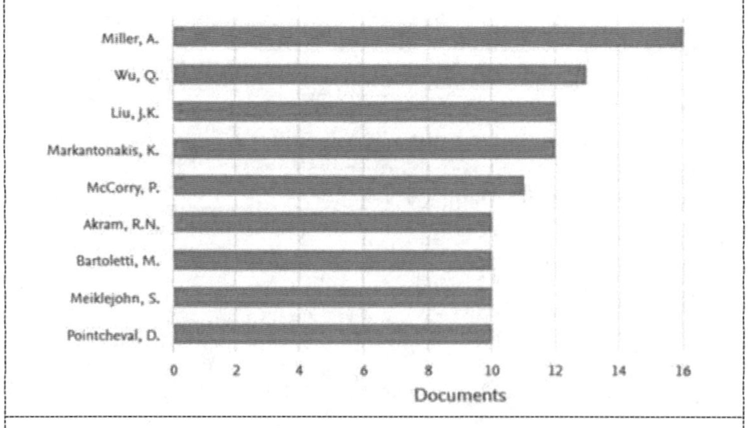

Figure 2. Most individual E-Money Literature Researcher

The researcher in the area of e-money to the most writings was Miller, A. with 16 documents with it. Pursued by Wu, Q. with 13 documents, Liu, J.K. with 12 documents, Markantonakis, K. with 12 documents, McCorry, P. with 11 documents, Akram, R.N. with ten documents, Bartoletti, M. with ten documents, Meiklejohn, S. with ten documents, Pointcheval, D. with ten documents, Biryukov, A. with nine documents and Bonneau, J. with nine documents.

3.3. E-Money Literature Most Common Nation

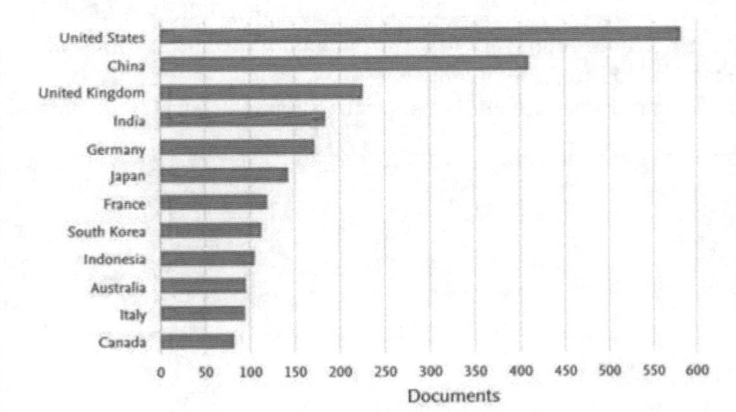

Figure 3. Nation Number of Annual Literature of E-Money

In e-money literature, the United States 582 academic documents was the leading research nation. Then, with 409 articles, the China followed. Then United Kingdom, India, Germany, Japan, France, South Korea, Indonesia, Australia, Italy, Canada and Taiwan.

3.4. The Largest Frequency of Literature of E-Money by Subject Area

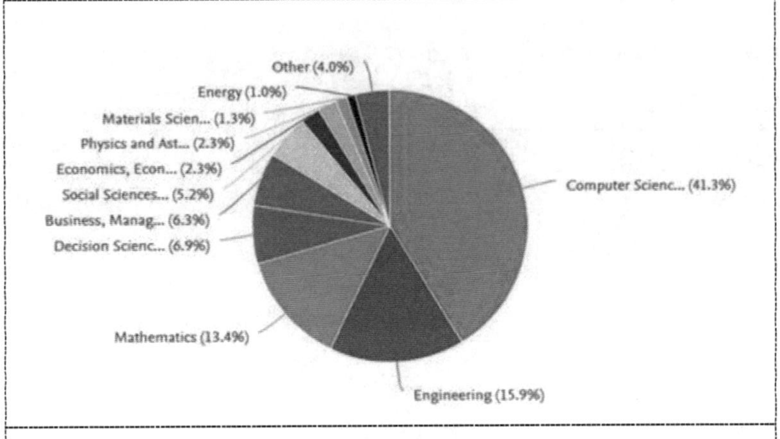

Figure 4. The Largest Frequency of Literature of E-Money by Subject Area

With 2424 documents (41.3 percent), computer science was the most frequent topic in international e-money literature in the subject field. Then it was followed by engineering with 931 papers (15.9%), mathematics with 787 papers (13.4%), Decision-science with 406 papers (6.9%), business, management and accounting with 371 papers (6.3%), social science with 307 papers (5.2%), economics, econometrics and finance with 137 papers (2.3%), physics and astronomy with 133 papers (2.3%).

3.5. Year Documents of E-Money Literature Sources

Lecture Notes in Computer Science Including Subseries Lecture Notes in Artificial Intelligence and Lecture Notes in Bioinformatics are the leader in the annual number of sources of e-money literature with 493 papers. Then, accompanied by 88 documents from the ACM International Conference Procedural Collection, Proceedings of the 38-document ACM Conference on Computer and Communications Stability, 37-document Physica A Statistical Mechanics and Its Applications, 33-document IEEE Access and 30-document Advances in Smart Systems and Computing.

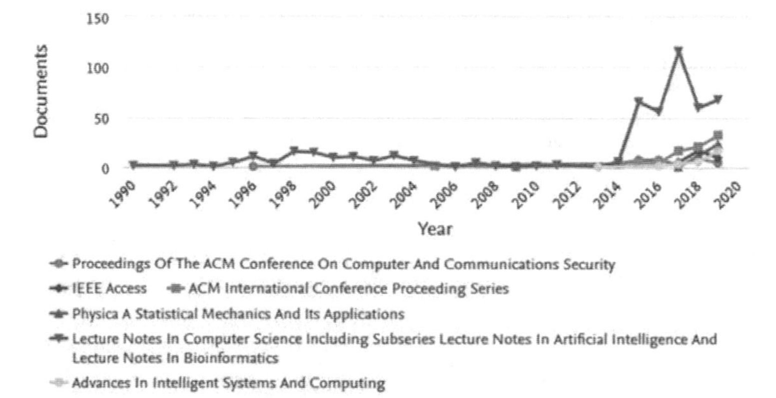

Figure 5. Number of Annual Documents Based on the E-Money Literature Sources

3.6. Per Year Documents of the E-Money Literature Sources

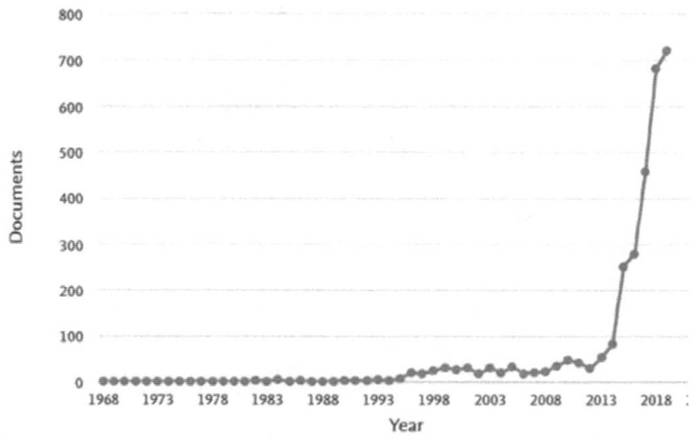

Figure 6. Number of Documents in the E-Money Overview Each Year

The number of international publications on the e-money Summary has seen a growing trend each year. This can be seen in Figure 6, the highest publication peak in 2019, with 720 papers. Work on e-money has been running since 1968. In e-money reports, the number of documents per year is 720 documents in 2019, 681 documents in 2018, 458 documents in 2017, 279 papers in 2016 and 250 papers in 2015.

3.7. The Document Cited of E-Money Literature

The study of Schierz, P.G., Schilke, O., Wirtz, B.W. was the most widely cited literature, in 2010 entitled "Understanding Consumer Acceptance of Mobile Payment Services: An Empirical Analysis" in "Journal of Electronic Commerce Research and Application" with cited by 447 documents. Then, followed by Tschorsch, F., Scheuermann, B. in 2016 entitled "Bitcoin and beyond: A technical survey on decentrelized digital currenvies" in "IEEE Communications Surveys and Tutorials" with cited by 443 documents and Luu, L., Chu, D.-H., Olickel, H., Saxena, P., Hobor, A. in 2016 entitled "Making smart contracts smarter open access" in "Proceedings of the ACM Conference on Computer and Communications Security" with cited by 413.

3.8. Literature Theme Map

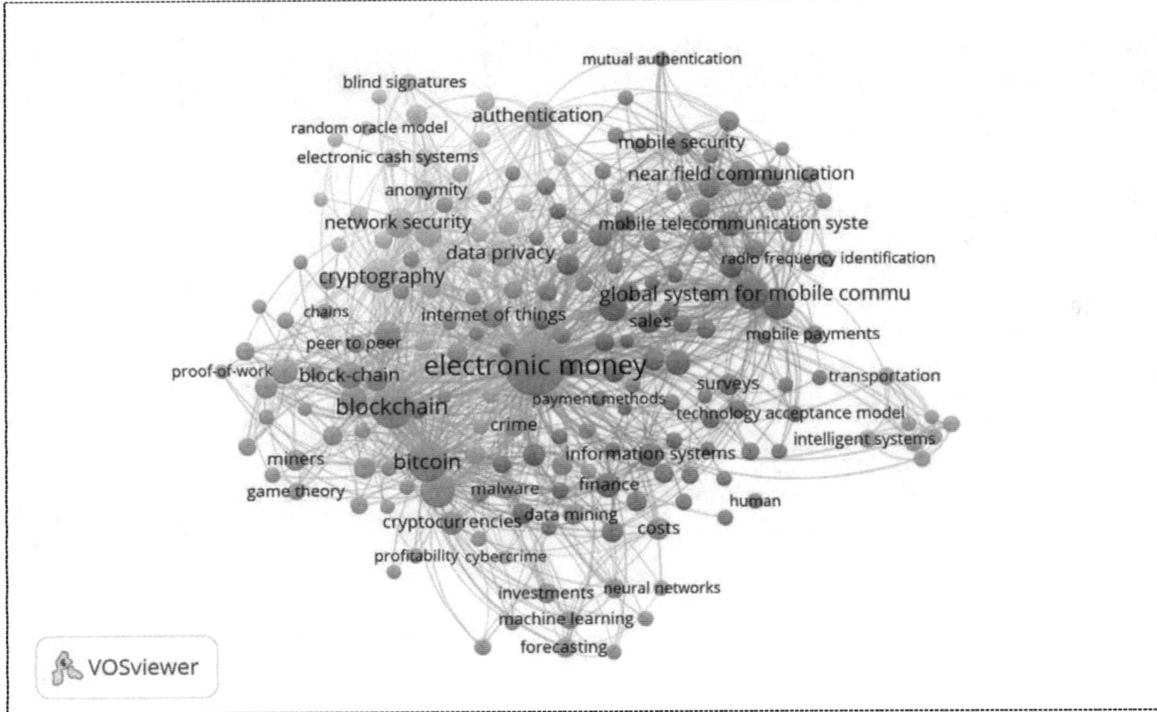

Figure 7. Network of Keyword

With analysis and visualization of the VOSViewer program, construction was developed on the e-money keyword framework for the e-money of literature theme map. Three repetitions were the criterion for the minimum amount of keyword-related documents. Therefore, 13,229 keywords among 197 keywords reached the thresholds.

From Figure 7. there were six literature theme groups dependent on study keywords regarding the international academic literature of e-money, simplified as well as abbreviated as MEFPII themes.

3.8.1. Mobile telecommunication system cluster (blue)
We can find mobile telecommunication themes in this cluster. The mobile security, mutual authentication, near field communication, and radio frequency identification keywords were connected to this cluster.

3.8.2. Electronic cash system cluster (yellow)
The keywords of network security, data privacy, random oracle model, blind signatures, authentication and cryptography dominated this cluster. Many of these keywords are linked to themes in electronic cash system.

3.8.3. Finance clusters (red)
We can find finance themes in this cluster. The mobile payment, payment method, information system, technology acceptance model, human and sales keywords were connected to this cluster.

3.8.4. Profitability cluster (green)
The keywords of profitability, block-chain, bitcoin, internet of things, game theory, proof-of-work, chains and miners dominated this cluster. Many of these keywords are linked to themes in profitability.

3.8.5. Investments cluster (purple)
We can find investments themes in this cluster. The costs, neural network, machine learning, forecasting, data mining and cryptocurrencies keywords were connected to this cluster.

3.8.6. Intelligent system cluster (orange)
The keywords of intelligent system and transportation dominated this cluster. Many of these keywords are linked to themes in intelligent system.

3.9. Network of Authorship

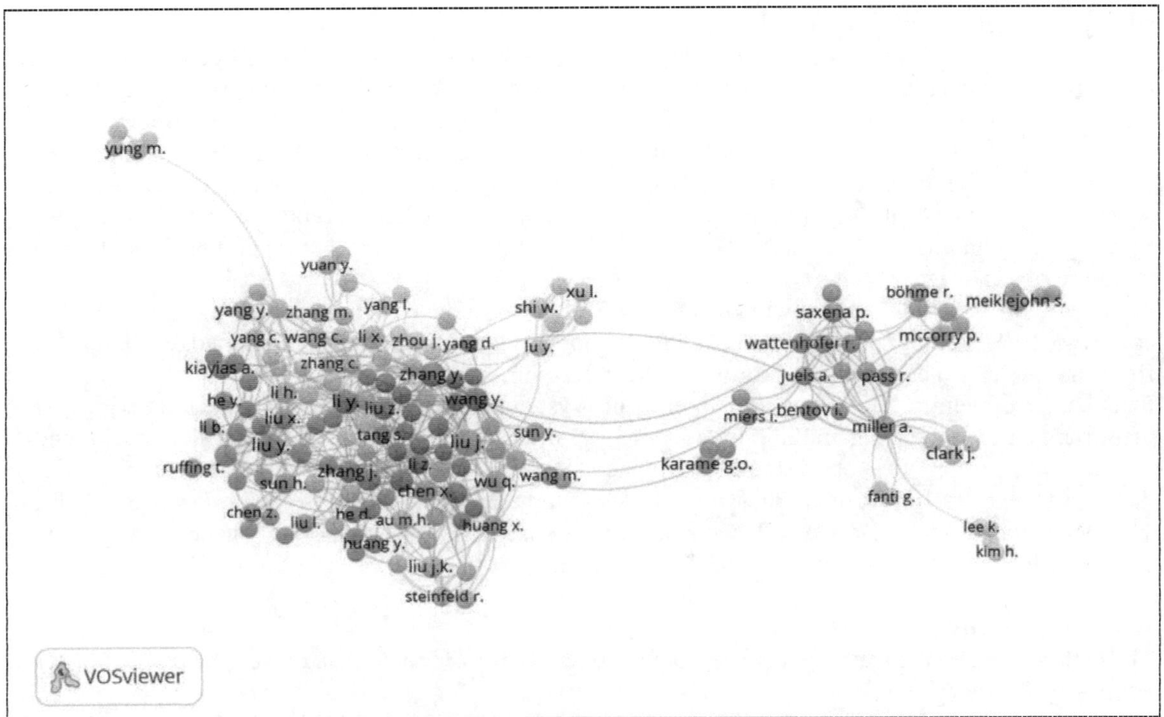

Figure 8. Authorship Network Map

With the VOSViewer program, construction was developed on the e-money researcher framework for the authorship network map. One document was one of the requirements for the minimum collection of publications per author. Thus, out of 14,125 researchers, 102 researchers who reached the thresholds were recognized.

As shown in the figure 8, there were nine group partnership networks between international researchers in e-money publications. The green cluster of e-money publication which contains Saxena, P., Wattenhofer, R., Juels, A., Pass, R., Bentov, I., Miller, A., Miers, I., Karame, G.O. They dominant from United States. The purple cluster which contains Böhme, R., Meiklejohn, S., Mccorry, P., Liu, I., Chen, Z., Sun, H., Liu, X., Liu Y., Ruffing, T. They dominant from United States. The yellow cluster which contains Yuan, Y., Zhang, M., Wang, C. They both from China. The pink cluster which contains Yang, Y., Yang, C., Li, X., Li, H. They both from China. The brown cluster which contains Kiayias, A., He, Y., Li, B., Li, Y. They both from England. The dark blue cluster which contains Yang, D., Zhang, Y., Liu, Z., Tang, S., Zhang, J., He, D., Huang, Y. They dominant from China. The light blue cluster which contains Au, M. H., Liu, J.K., Steinfeld, R. They dominant from United States. The oorange cluster which contains Sun, Y., Wang, M., Liu, J., Chen, X., Wu, Q. They dominant from China. Then the red cluster which contains Wang, Y., Li, Z., Huang, X. They dominant from China.

4. Conclusion

The results of this research revealed that there is an annual trend towards a spike in the amount of international literature on e-money, there were maps and visual patterns. With 582 papers, the United States was the country with the greatest contribution to literature in e-money studies. In the publication of the e-money literature, Chinese Academy of Sciences was the most active research institution with 41 papers. In the e-money literature, the individual academic researcher with the most prolific publications was Miller, A. 16 papers with it. With 2424 documents (41.3 percent), the most intensively studied areas published in the e-money literature were computer science. The "Lecture Notes in Computer Science Including Subseries Lecture Notes in Artificial Intelligence and Lecture Notes in Bioinformatics" with 493 documents was the majority of annual documents by the source in the e-money literature. With 720 papers, the highest publication of worldwide scholarly literature in e-money studies was in 2019. The works of Schierz, P.G., Schilke, O., Wirtz, B.W. were mostly literature with the most citations. In 2010, the "Journal of Electronic Commerce Research and Application" cited 447 documents entitled " Understanding Consumer Acceptance of Mobile Payment Services: An Empirical Analysis". There were nine researcher partnership groups linked to the literature of e-money.

In terms of contributing knowledge implications, this study recommends a classification of the convergence axis comprising of literature in e-money to classify the body of knowledge created from fifty-one years of academic literature: Mobile telecommunication system, Electronic cash system, Finance, Profitability, Investments, Intelligent system, abbreviated as MEFPII themes. The identification of key themes in the e-money area, as practical implication, contributes to an awareness of the creation of practical studies to clarify general contexts and topics, as well as research gaps. All this will lead to fresh research addressing a lack of study and specialized expertise in the disciplines. The most studied themes often reflect the ability to contribute of e-money to finance, e-commerce, digital business, investments and electronic cash system.

Acknowledgments
The authors are indebted and thanks Airlangga University for access to data on ademic publications on Scopus.

References

[1] O. E. Pyrkina and S. A. Zadadaev, "Application of Crash-Tests for Graph Analytics of Electronic Payment Complex System for Investigation of System Economic Security," in 2019 Twelfth International Conference "Management of large-scale system development" (MLSD), 2019, pp. 1–4, doi: 10.1109/MLSD.2019.8911069.
[2] N. Polyakov, F. Tishkin, L. Cherckesova, and O. Safaryan, "Security of User Authentication in \ Payment Systems in the Agricultural Value Chain," IOP Conf. Ser. Earth Environ. Sci., vol. 403, p. 012161, Dec. 2019, doi: 10.1088/1755-1315/403/1/012161.
[3] C. Khiewngamdee and H. Yan, "The Role of Fintech E-payment on APEC Economic Development," J. Phys. Conf. Ser., vol. 1324, p. 012099, Oct. 2019, doi: 10.1088/1742-6596/1324/1/012099.
[4] Z. Bezhovski, "The Future of the Mobile Payment as Electronic Payment System," Eur. J. Bus. Manag., vol. 8, no. 8, pp. 127–132, 2016.
[5] P. P. Patil, Y. K. Dwivedi, and N. P. Rana, "Digital Payments Adoption: An Analysis of Literature," 2017, pp. 61–70.
[6] S. Hapugoda and C. Hewagamage, "Analysis of the Necessity of Developing Regulatory Guidelines for E-Money Operations in Sri Lanka," in 2015 Fifteenth International Conference on Advances in ICT for Emerging Regions (ICTer), 2015, pp. 272–272, doi: 10.1109/ICTER.2015.7377705.
[7] H. Mandari and D. Koloseni, "Why Mobile Money Users Keep Increasing? Investigating the Continuance Usage of Mobile Money Services in Tanzania," J. Int. Technol. Inf. Manag., vol. 26, no. 2, pp. 77–86, 2017.
[8] V. V. Pshenichnikov and A. V. Babkin, "Digital Money as A Product of the Vevelopment of Information and Telecommunication Technologies," in 2017 International Conference "Quality Management,Transport and Information Security, Information Technologies" (IT&QM&IS), 2017, pp. 267–273, doi: 10.1109/ITMQIS.2017.8085808.
[9] R. Agrawal, "Mobile Money Empowering People Living at Bottom of Pyramid and Boosting

Socio-Economic Development in a Big Way," Econ. Anal., vol. 49, no. 1–2, pp. 15–23, 2016.

[10] A. Seetharama and J. Rudolph Ra, "An Empirical Study on the Impact of Earnings per Share on Stock Prices of a Listed Bank in Malaysia," Int. J. Appl. Econ. Financ., vol. 5, no. 2, pp. 114–126, Feb. 2011, doi: 10.3923/ijaef.2011.114.126.

[11] S. M. T. Toapanta, A. A. S. Balladares, D. F. H. Subia, and L. E. M. Gallegos, "Prototype of a Security Model to Mitigate Risks in the Management of Electronic Money in Ecuador," in 2019 Third World Conference on Smart Trends in Systems Security and Sustainablity (WorldS4), 2019, pp. 87–93, doi: 10.1109/WorldS4.2019.8903959.

[12] S. Schuh, "Measuring Consumer Expenditures with Payment Diaries," Econ. Inq., vol. 56, no. 1, pp. 13–49, 2017.

[13] D. Haryadi, Harisno, V. H. Kusumawardhana, and H. L. H. S. Warnars, "The Implementation of E-money in Mobile Phone: A Case Study at PT Bank KEB Hana," in 2018 Indonesian Association for Pattern Recognition International Conference (INAPR), 2018, pp. 202–206, doi: 10.1109/INAPR.2018.8627055.

[14] S. A. E. A.Elmagid and A. Z. Karrar, "E-payment Adoption by Petroleum Companies in Sudan," in 2019 International Conference on Computer, Control, Electrical, and Electronics Engineering (ICCCEEE), 2019, pp. 1–8, doi: 10.1109/ICCCEEE46830.2019.9071021.

[15] D. W. C. Soon, T. C. Wei, and L. W. Lee, "Factors Influencing Consumers' Perception on Mobile Payment among Generation Z in Malaysia," in Proceedings of the 2nd International Conference on Big Data Technologies - ICBDT2019, 2019, pp. 316–319, doi: 10.1145/3358528.3358532.

[16] A. Purnomo and M. Firdaus, "E-Money Research Dataset (1968-2019)," Mendeley Data, 2020. https://data.mendeley.com/datasets/4pc7mnmjdy/1.

[17] N. J. van Eck and L. Waltman, "Software survey: VOSviewer, A Computer Program for Bibliometric Mapping," Scientometrics, vol. 84, no. 2, pp. 523–538, Aug. 2010, doi: 10.1007/s11192-009-0146-3.

[18] A. Purnomo, T. Susanti, A. K. Sari, M. Firdaus, and R. Dewi, "A Study of Digital Entrepreneurship through Bibliometric Visualizing from 1993 to 2019," in 2020 International Conference on Information Management and Technology (ICIMTech), 2020, pp. 911–915, doi: 10.1109/ICIMTech50083.2020.9211270.

[19] A. Purnomo, A. Septianto, D. U. Sutiksno, M. I. Hikmawan, and R. D. Kumalasari, "Technopreneur Publication: A Bibliometric Analysis (2000–2019)," in 2020 International Conference on Information Management and Technology (ICIMTech), 2020, pp. 521–526, doi: 10.1109/ICIMTech50083.2020.9211111.

[20] B. Ranjbar-Sahraei and R. R. Negenborn, Research Positioning & Trend Identification. Walanda: TU Delft, 2017

IOP Conf. Series: Materials Science and Engineering 1071 (2021) 012029 doi:10.1088/1757-899X/1071/1/012029

Systematics Review on the Application of Social Media Analytics for Detecting Radical and Extremist Group

R T Adek[1], Bustami[2] and M Ula[3*]
[1,2]Informatics Department, Universitas Malikussaleh, Lhoksumawe
[3]Information System Department, Universitas Malikussaleh, Lhoksumawe

E-mail: munirulula@unimal.ac.id

Abstract. Recently, social media platforms such as Twitter, Tumblr, Facebook, YouTube, blogs and discussion forums are being mistreated by radical groups to promote their ideologies and encourage radicalization. Social medias also have been used to create online extremist community and recruit new followers. In this paper work, the authors conduct a schematics literatures review on all available techniques and perform a comprehensive analysis on the application of social media analytic for detecting radical group to understand the circumstances, trends and its gaps. Further, the author provides the characterization, classification and meta-analysis in order to achieve a better understanding of the literature on the extremist detection through intelligent social media. It is found that for over the last 10 years many researchers have been conduct deep investigation on the use of social media analytics on predicting and identifying online radicalization. Besides, data source, features, geolocation, language, machine learning techniques, and tools have been applied on the recent literatures to detect those cyber-extremist activists. This paper also highlighting the performance measurement methods that have been used by researcher for detecting extremist group and radical communities. The goal of this research is to provide an academic base for ongoing research in the developing machine learning algorithm for detecting extremist and radical contents in social medias.

1. Introduction

Over the last decade, social media has emerged into a dynamic form of interpersonal communication around the world. It facilitates users to constantly share information and continuously connect and convey opinion to the world through medias. For example - social networking (Facebook), micro-blogging (Twitter, Tumblr), image sharing (Imgur, Flickr) and video hosting and sharing (YouTube, Dailymotion, Vimeo) [1]. Likewise, the simplicity of navigation, low barriers to publication (users only need to have a valid account on the website) and anonymity (freedom to upload any content without revealing their real identity) have caused users to abuse this website in several ways by uploading offensively. and illegal data [1][2]. Further, popular social media websites, blogs, forums are often misused by divers of groups to promote online radicalization (also referred to as cyber-extremists, cyber-crime and cyber hate propaganda radicalization (also referred to as cyber-extremists, cyber-crime and cyber hate propaganda).

Previous research showed that extremist groups often delivered a hate speech, offensive comments and messages that focused on their mission. a large number of hate groups use popular social media websites to disseminate their ideology with extremist content for their readers [2] [3]. They communicate with other existing groups and form virtual communities on social media and sharing a common agenda. They use social networks as a platform to hire new members by gradually reaching a worldwide audience that helps persuade others to commit violence and terrorism [4][5]. Researchers from various disciplines such as psychology, social science and computer science are constantly developing tools and knowledge on fighting technique and these problems online radicalization.

Hence, automatic detection and radicalizing content analysis is an important issue in the research domain of intelligent Security Informatic. Monitoring the presence of such content on social media and this issue in real time is predominant for security analysts who works on law enforcement agencies.

Since 2005, for more than 10 years, Information Security Intelligent (ISI) [5] [6] [7] has been proposed number of approaches, techniques, algorithms and tools to mine the data and provide solutions on emerging problems. The aim of the research in this journal is to carry out a systematic literature survey of the previous techniques as documented in scientific articles. Accordingly, the author's goal is to conduct a comprehensive analysis of those articles in order to get better understanding on research gap, techniques and future recommendation.

2. Research Method

This systematic literature review has been done by applying 6 stages process for a systematic review research method [8]. The process was started by research question formulation. The second stage was to determine the required characteristic for the study. The third stage was collecting relevant papers from the research databases. The fourth stage was selecting related research papers and cleansing the irrelevant papers. The fifth stage was synthesizing the pertinent information from the literatures and the last stage was tabulation and report writing [9].

To provide guideline of the systematics review, the following research question was addressed:

1. What data sources and features are used in detecting radical and extremist group?
2. What machine learning techniques are used in detecting radical and extremist group?
3. What performance evaluation method are used in detecting radical and extremist group?

2.1. Systematic Review Process

The data collection process in this systematic review is as shown in Figure 1. The research papers for literature review were retrieved from many credible online databases dan google scholar. This research utilize Google Scholar is because it provides more articles coverage and index and also offer a strong mechanism to explore related works, ideas and authors [9] [10]. Besides, Google Scholar also produces data on how often and how recently journals have been cited which is also a useful metric (assessing impact) for conducting author literature surveys. Variation of keywords used in search engines and online databases for data collection are “Extremist Detection, Radical Detection, social media, machine learning, 'online community radicalization detection”. After the research articles were collected. Title, Keywords and Abstracts were analysed. A meta-analysis process was conducted on the article and determines the relevance of the articles to the author's topic or focus area. The author conducts a one-class classification (Rule Based Classifier) on each article and checks whether it meets the specified scope and focus.

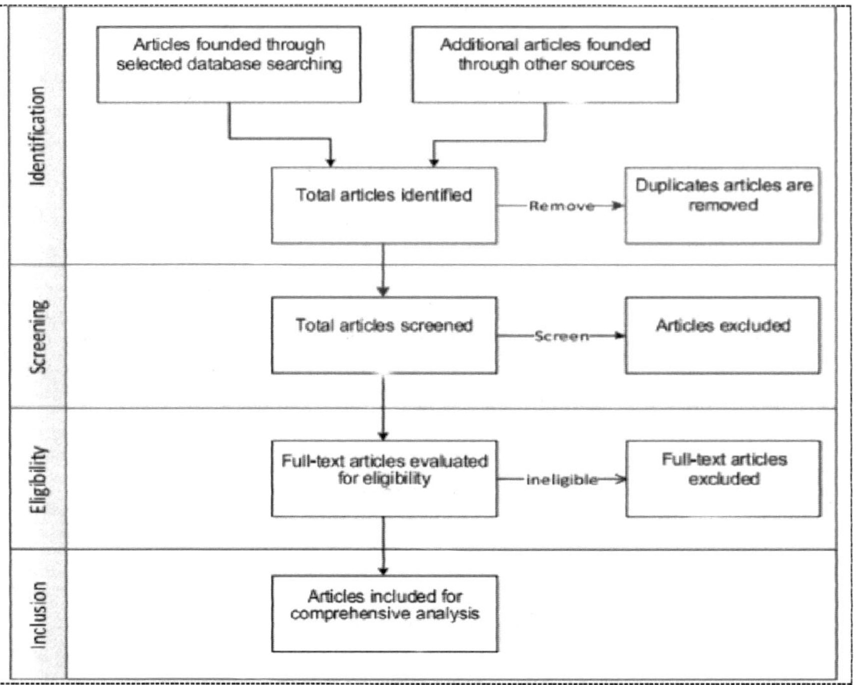

Figure 1. Systematic review process

3. Systematic Review Result

3.1. Data Source

From 2010 to 2020, there are more than 40 publications on online extremism detection. This review has identified many publications using multiple data sources and experimental datasets over the last decade. Figure 2 reveals that there is some work on all popular social networking platforms, micro-blogging websites [8] [9], video sharing [10] [11] [12] [13] and image hosting websites [14]. Other researchers have used blogs [15], forums [16], web documents, news articles [17] and websites [18] for hate and extremism detection. The figure shows that most of the research publications using social media platform as their data source. The most famous social media website is twitter. Because twitter offer easy data scrapping using API for developer. Other articles have utilizing blogs, discussion forums, news articles and websites as data sources. However, YouTube, also gradually become popular data source for researcher for detecting extremist and radical contents. Moreover, the preceding research indicate that YouTube is the most prevalent platform used for hate promotion [19].

Figure 2 shows that there are 28% of research papers have used Twitter, 24% has used multi data source, 19% has used Facebook, and 21% YouTube as a data source. The pie chart above shows that Twitter, Facebook, YouTube and multiple sources have been being used as data source by many researchers for detecting extremist and radical communities for the last decade [11] [12] [20] [21]. Twitter as a famous micro-blogging website is very popular platforms for radical group and community. It has been extensively used by the extremist to deliver their messages and propaganda to their communities. However, many researchers have used multiple platforms as their data sources for detecting extremist and radical communities. These multiple sources are including Facebook, Instagram, Twitter and others. Other data sources like news, blog, and forum have gradually received less attention from researchers in the last decade [22] [23] [24].

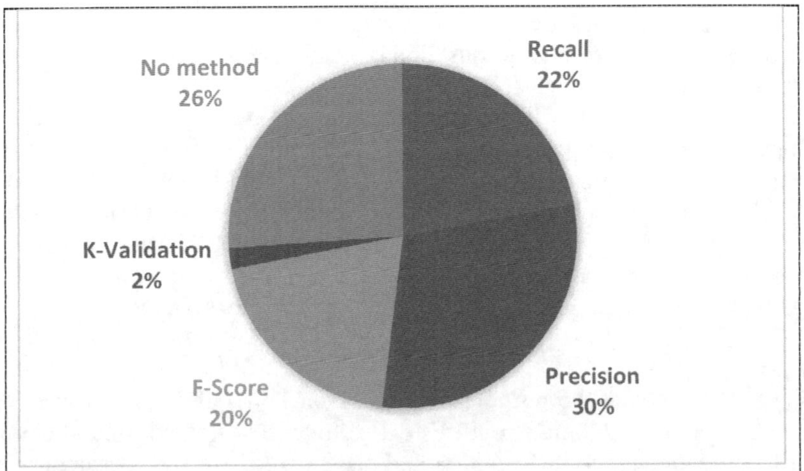

Figure 2. Data Source

3.2. Data Features

This review classifies research articles based on three discriminatory features as follows:

1. Text;
   Text defines as the contextual metadata of a post, tweet content, video title, hashtags in the video description, etc.
2. Link
   Link defines as the connection between two user account. It could be a "repost blog by" on Tumblr, "followers" on Twitter, "liked videos" on YouTube, etc.
3. Demographic information
   This information is a metadata based on post statistics and user profiles. For example, number of comments on videos, number of favourites on tweets, notes on Tumblr posts etc.

The systematics review results show that; 16 papers using text as their data source, 24 papers use links as their data, and 28 research papers use demographic information as their data. Furthermore, this review records that 25% of research papers use all three features in their analysis, while only 30% of these papers use text and demographic information to detect the extremist group and radical communities. Only 5% of the reviewed papers, that only use link as their data for detecting radical contents and extremist groups.

3.3. Type of Contents

In this review, type of contents in research articles are classified into 3 sub-categories namely 'content', 'user profile' and 'community' based on the type of content analyzed to provide solutions to fight and combat online extremism. Results of literature analysis discovers that out of 36 articles, in 35 papers, the authors conducted experiments on contextual metadata while in 18 papers, the authors extracted user profile information. Likewise, in 16 articles, they mine the profiles of linked users and their communities. Content analysis is an important feature for identifying extremist content.

3.4. Language

Social medias contain multi-lingual text in metadata, therefore, in this review, authors categorize research papers into three language classification. The review results reveal that among the 36 research articles, 14 articles are successfully detect the extremist group and radical content posted in Arabic. Furthermore, among these 14 articles, the methodology applied in 7 articles is able to detect

IOP Conf. Series: Materials Science and Engineering 1071 (2021) 012029 doi:10.1088/1757-899X/1071/1/012029

other non-English texts as well. The review also exposed that all the machine learning technique used in the past research were able to detect extremist and radical contents posted in English.

3.5. Geolocation
Several research papers on detecting extremist group and radical communities have been focused on the metadata determined and originates from a specific region and country. Therefore, in this review, authors have classified 36 research papers and articles into three categories which are Domestic, International, and Others. The review result shows that there are 20 research papers and articles that focus on detecting extremist and radical group taking place worldwide and other 5 articles focused on US domestic region and 9 research papers focus on middle east and Asian region.

3.6. Machine Learning Techniques
The graph in Figure 3 shows that Support Vector Machine (SVM) and Naïve Bayes (NB) are among the famous machine learning techniques used for detecting extremist and radical communities in last decade [25] [26]. Other machine learning techniques also used by researchers are KNN text classification, Rule Based Classifier, Decision Tree, Clustering, Exploratory Data Analysis (EDA), Topical Crawler are the most widely used techniques for online radicalization detection on social media websites [27] [28] [29]. Twitter, Facebook, Instagram and YouTube are users generated content websites. Those websites generated terabyte multimedia data in a month. Therefore, text classification techniques like EDA and KBF algorithms are very well and commonly used to identify extremist content on social media [21] [28].

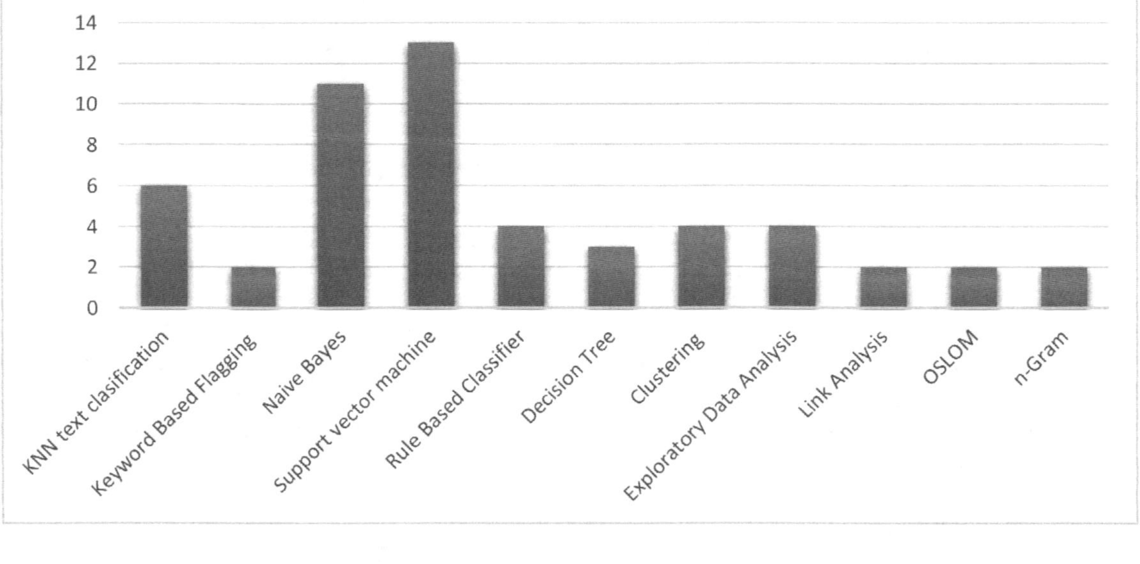

Figure 3. Machine learning techniques used by researchers

Topical crawler and link analysis is a technique used to enter social media websites through navigational links and identify similar users and find the hidden communities [11] [15] [22]. The crawler topical is a recursive process that adds and removes nodes after each iteration. Starting from the seed node, traverse in the graph navigating through multiple links and returning all relevant nodes to the given topic. Breadth First Search, Depth First Search and Best First Search are techniques to select neighbours and navigate them through external links. These links and neighbours are varied for different social networking websites. For example, if the user (u) posts a tweet (t) then the neighbours can become your users' followers, or the user likes and tweet or re-blogs (in the case of Tumblr) that post. Likewise, on YouTube, if your user posts v video then the link can redirect back to the channel of the user who subscribed or posted a comment on video v. N-gram is another technique for classifying textual data in detecting radical contents based on some discriminatory features [4].

Other machine learning technique for detecting extremist is OSLOM (Order Statistics Local Optimization Method). It is one of the clustering methods designed for detecting graphs and social media networks. OSLOM has been applied for detecting extremist and radical group network [23] [24].

3.7. Performance Evaluation Method

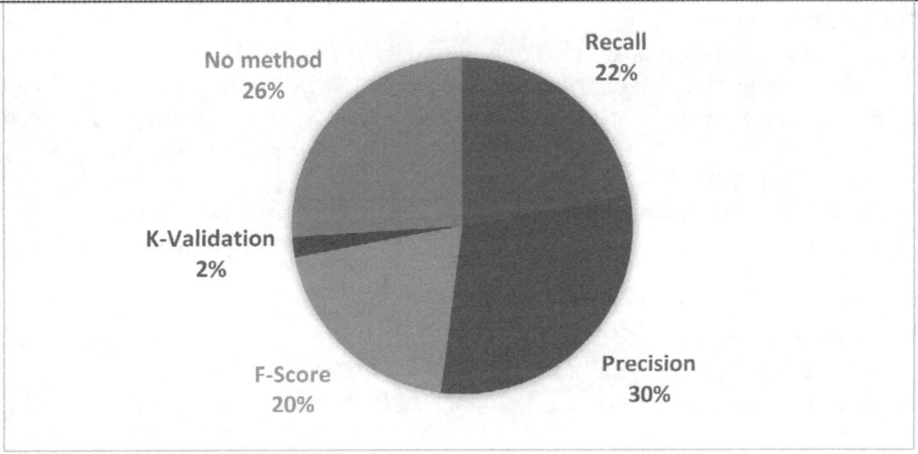

Figure 4. Performance Measurement Method

In this review we also analyse the performance measurement method used by many researchers. This section purposes are to provide answer for the last research question; What performance evaluation method are used in detecting radical and extremist group? The result of this systematics review shows that Precision, Retrieval, K-cross validation are commonly used to measure the performance of the machine learning technique used. In this work, author classify performance evaluation method into 5 sub-categories which are F-score, Precision, Retrieval, K-cross validation. F-score is the average harmonic of Precision and Recall while Accuracy calculates the "correctness" of the proposed method. *Figure 4* shows the distribution of the number of articles using various measures to examine the performance proposed over the last decade. Figure 4 also reveals that there is only one article using K-cross validation for evaluation in 7 articles no evaluation method is mentioned. The authors also consider that although 50% of the articles analysed only textual data to identify extremist content, only 30% of the articles used precision, while only 20% of the articles used recall as a measure for evaluation.

4. Conclusion

Applying intelligent in social media to identify online radicalization is an area that has attracted the attention of several researchers over the past 10 years. Characterization and meta-analysis conducted on existing studies on online radicalization detection revealed that identifying the presence of metadata based on extreme contextual content is the most commonly used feature. However, demographic and activity information from user profiles and links between two users are discriminatory features for finding extremists. The author observes that many existing techniques are capable of mining multi-lingual texts such as Arabic and capturing relevant information. The author also observed that precision is the most commonly used evaluation method to test the effectiveness of the results in community detection and identification of extremism content, and Social Network analysis.

References

[1] Ahmad S, Asghar M Z, Alotaibi F M, and Awan I, 2019. Detection and classification of social media-based extremist afliations using sentiment analysis techniques, Human Centric Computing Information Science, (2019) 9:24, https://doi.org/10.1186/s13673-019-0185-6.
[2] Drus Z, Khalid H 2019, Sentiment Analysis in Social Media and Its Application: Systematic Literature Review. Procedia, The Fifth Information Systems International Conference 2019
[3] Sharif O, Hoque M M, Kayes A. S. M., Nowrozy R, and Sarker H I, 2020, Detecting Suspicious Texts Using Machine Learning Techniques. MDPI - Applied sciences Article 2020.
[4] Garg U and Kaur S, 2020. Systematic Review on Machine Learning Approaches for Sentiment Analysis. International Journal of Advance Science and Technology Vol. 29, No. 10S, (2020), pp. 2760-2764
[5] Arunachalam, R. dan Sarkar, S. 2013., 'The new eye of government: Citizen sentiment analysis in social media'. IJCNLP 2013 Workshop on Natural Language Processing for Social Media (SocialNLP) hal. 23–28.
[6] BBC. 2016. MIT dan JAT: Dua Kelompok Teror Indonesia Terkait ISIS. Diunduh dari www.bbc.com/indonesia/berita_indonesia/2016/01/160115_indonesia_explainer_kelompok_teror, accessed on 17 May 2019.
[7] Manning D C, Surdeanu M, Bauer J, Finkel J, Bethard S. J, and McClosky D. 2014. The Stanford CoreNLP natural language processing toolkit. In Proceedings of 52nd Annual Meeting of the Association for Computational Linguistics: System Demonstrations, pages 55–60, Baltimore, Maryland, June 2014. Association for Computational Linguistics
[8] Christian D F, Kembro J, and Andreas. 2017 A New Paradigm for Systematic Literature Reviews in Supply Chain Management. Journal of Supply Chain Management Wieland 53 (4): 67-85
[9] Liu H, Christiansen T, Baumgartner Jr W A, and Verspoor K. 2012 Biolemmatizer: a lemmatization tool for morphological processing of biomedical text. J. Biomedical Semantics, 3(3):17, 2012.
[10] Hashemi, Mahmood. S, Jingsha He. 2017. LA-Based Approach for IoT Security. Journal of Robotics, Networking and Artificial Life, Vol. 3, No. 4 (March 2017), 240-248.
[11] IHS Jane's Defence Weekly. (2012). Annual Defence Report: Middle East and Africa, 7 December 2012.
[12] Liu, B. (2012). Sentiment Analysis And Opinion Mining: Morgan dan Claypool Publisher.
[13] Pak, A. dan Paroubek, P. (2010). 'Twitter as a corpus for sentiment analysis and opinion mining'.Proceedings of the Seventh conference on International Language Resources and Evaluation (LREC'10) hal. 1320–1326.
[15] Selvam, B. dan Abiram, S. (2013). A survey on opinion mining framework.
[16] Agarwal S, Sureka A (2015) Topical Crawler for Uncovering Hidden Communities of Extremist Micro-Bloggers on Tumblr, 5th Workshop on Making Sense of Microposts, Microposts 2015, May 18th, 2015, Florence, Italy.
[17] Agarwal S, and Sureka A, Using KNN and SVM Based One-Class Classifier for Detecting Online Radicalization on Twitter, ICDCIT 2015, Springer International Publishing Switzerland 2015, LNCS 8956, pp. 431–442, 2015.
[18] Agarwal A, and Sureka A, 2015 Topic-Specific YouTube Crawling to Detect Online Radicalization DNIS 2015, Springer International Publishing Switzerland 2015, LNCS 8999, pp. 133–151, 2015.

[19] Agarwal A, and Sureka A, 2014, A Focused Crawler for Mining Hate and Extremism Promoting Videos on YouTube, HT'14, ACM, September 1–4, 2014, Santiago, Chile. 978-1-4503-2954- 5/14/09.

[20] Scanlon J R, and Gerber M S, 2014 Automatic detection of cyber-recruitment by violent extremists, Scanlon and Gerber Security Informatics 2014, 2014 Scanlon and Gerber; licensee Springer.

[21] Chen F, Daniel B. Neill, 2014 Non-Parametric Scan Statistics for Event Detection and Forecasting in Heterogeneous Social Media Graphs, KDD'14, 2014 ACM, August 24–27, 2014, New York, NY, USA.Copyright 978-1-4503-2956-9/14/08

[22] Jiejun Xu, Tsai-Ching Lu, Compton R, and Allen D, 2014 Civil Unrest Prediction: A Tumblr-Based Exploration, Springer International Publishing Switzerland 2014, SBP 2014, LNCS 8393, pp. 403–411, 2014.

[23] D. O'Callaghan, D. Greene, M. Conway, J. Carthy, and P. Cunningham. Uncovering the wider structure of extreme right communities spanning popular online networks. arXiv, preprint arXiv:1302.1726, 2013.

[24] Wadhwa P. and Bhatia M, 2013 Tracking on-line radicalization using investigative data mining. In Communications (NCC), 2013 National Conference on, pages 1{5. IEEE, 2013.

[25] I.-H. Ting, H.-M. Chi, J.-S. Wu, and S.-L. Wang. (2013) An approach for hate groups detection in facebook. The 3rd International Workshop on Intelligent Data Analysis and Management, Springer Proceedings in Complexity, pages 101{106. Springer Netherlands, 2013.

[26] Goodwin M, 2013 The Roots of Extremism: The English defence League and the Counter-Jihad Challenge. Chatham House.

[27] Patil G., Manwade K., and Landge M. P. 2013. A novel approach for recognized and overcrowding of terrorist websites. International Journal of Engineering Trends and Technology, Volume4Issue3-2013, ISSN: 2231-5381, pp 463-470

[28] Deshwal A.and Sharma S., 2016 Twitter sentiment analysis using various classification algorithms, 2016 5th International Conference on Reliability, Infocom Technologies and Optimization (Trends and Future Directions) (ICRITO).

[29] Ferrara E, Wang WQ, Varol O, Flammini A, Galstyan A (2016) Predicting online extremism, content adopters, and interaction reciprocity. International conference on social informatics. Springer, New York, pp 22–39

IOP Conf. Series: Materials Science and Engineering 1071 (2021) 012030 doi:10.1088/1757-899X/1071/1/012030

IOT-BASED REFRIGERATOR MONITORING SYSTEM

Mirza Khamali[1*] and Wendi Usino[1]
[1]Fakultas Teknologi Informasi, Universitas Budi Luhur Jl. Raya Ciledug, Petukangan Utara, Kebayoran Lama, Jakarta Selatan 12260

E-mail: mirzakhamali93@gmail.com

Abstract. Refrigerators are very valuable asset of beverage distribution companies and also assets that are lent to customers for the business process of selling beverage products. Because of the slow response in complaining about damage of the refrigerator from customers, so that refrigerator monitoring system is a technology that creates practical solution for monitoring the refrigerator. Integrated sensors will send data for monitoring through the internet network provided temperature, light, electric current and also can help to decide the division of tasks for technicians who will repair automatically. With a refrigerator monitoring system that uses IoT in beverage distribution companies, hopefully that a monitoring system in the refrigerator is accurate and fast with notifications using email that directly approved by technicians, providing temperature information in the refrigerator, refrigerator lights, making the schedule of the automation system which is directly submitted to the technician so that the repair process is faster.

1. Introduction

The background of this research is considering the numbers of customer's complains about the damage that happen in the refrigerators or the lamps of the refrigerators that are lent by PT. XYZ. Besides that, the slow fixing process because it is done only when customers report the issue caused the product selling and customer satisfaction are coming down.

Table 1. Top Five of Fixing Visits Duration after the Reports

Data Problem	Total	Average Time Response (Hour)
Asset/Products are not cold	19546	49.32321
Asset is totally off	5889	163.7117
The lamps are off/broken	5792	163.7117
Asset sounds noisy (not normal)	3073	313.7219
Electricity short circuit problem – OHS	2680	359.5036

In this case study of PT.XYZ, refrigerator is the valuable company asset that is lent to customers for the business process of selling drinking products. From the data analysis above, it needs development or creating system using IoT that will support monitoring process toward refrigerator, so that it can be done quickly and accurately [1].

Now, communication technology has imroved use of Internet of Thing (IoT). It is a technology that is able to connect all types of devices [2]. Internet of Things (IoT) offers various communication designs for handling process, monitoring and setting by using internet communication [3]. One of the utilizations of IoT is Refrigerator monitoring or smart refrigerator.

2. Research Method

Collecting data method of this research is using literature study by studying literatures, research journals and other sources that are related with this research. Researcher did observation using field study which is a technique of collecting data by going to the place by observing the problems directly in the venue systematically, incidents, behavior, objects that can be seen and other things that are needed to support the research. Interview is a data collection technique that is done through face to face and asking – answering directly between data collector towards the data source. From the data collection which is obtained by PT XYZ, data that will be lent is data one year in Jakarta area.

Analysis that is used is qualitative data analysis. Qualitative data analysis is done by collecting, selecting, grouping and taking note the data that is obtained from the source and also giving code so that the data source still can be searched. Monitoring system planning using IoT based, the researcher does integration between software and hardware that can be applied; it is waterfall method.

This application that the researcher develops is an embedded program application for wemos, web service for applications from hardware to send data, console application for fuzzy logic and web – based to monitor dashboard and notification. Hardware planning that the researcher conducts are planning of prototype diagram block tool and electronic module which is used in creating refrigerator monitoring prototype using IoT based.

After implementation step is done, the next step is testing the system that purpose to the final system result that is implemented as the needs. The testing of refrigerator monitoring system with IoT based (case study: PT. XYZ), the researcher uses black box testing. The testing is done only by running or executing the unit as the expected process. In this testing step by using black box testing, the researcher will test the system function, login, control system, authentication process, monitoring system and application setting. The research steps are asking permission for conducting the research, observing the research objects, collecting data process, the analysis method that will be used and project implementation.

3. Results and Discussion

3.1. Application Design

The application designed in this study was developed using the C, C #, data-based programming language using mysql and sql server. In this research, developing an application for monitoring the refrigerator that reports damage to the refrigerator temperature, refrigerator lights and the current connected to the refrigerator [4][5].

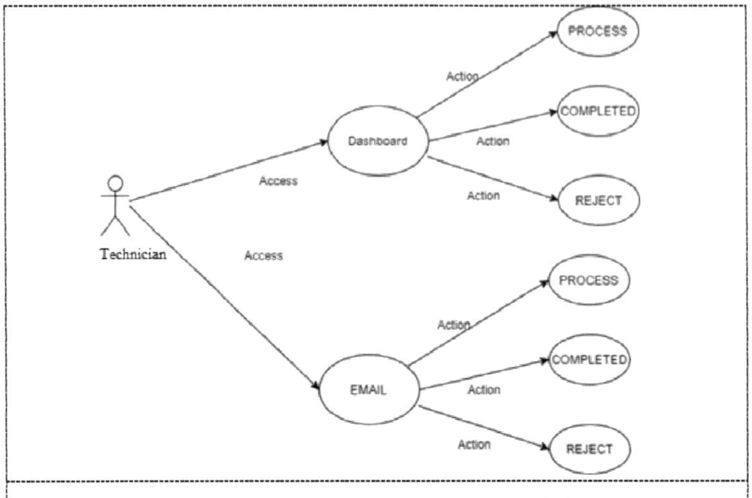

Figure 1. Use Case Diagram (Technician).

From the technician use case diagram image above, the process is described as follows:

Technicians can directly access the IoT-based refrigerator monitoring system application to see a list of refrigerators that will be processed for repair, after seeing the details of the request by the technician, the technician will take a process action by pressing the process, completed or reject button.

The technician will receive a notification from an email regarding a ticket request from a damaged refrigerator. The technician will click the link from the email to see the details of the request to perform the process action, namely by pressing the process, completed or reject button.

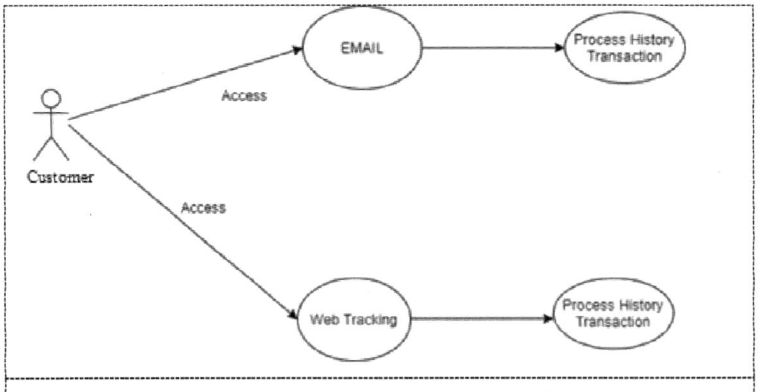

Figure 2. Use Case Diagram (Customer).

From the customer use case diagram above, the process is described as follows:

The customer will receive an automatic notification to an email containing the ticket number, then the customer will track the process via the link in the email. In addition to using email, subscribers can perform tracking from the web provided for tracking from the ticket number in the email.

The hardware design that the authors do includes designing a block diagram of the prototype tool, and the electronic module used in making IoT-based refrigerator monitoring prototypes.

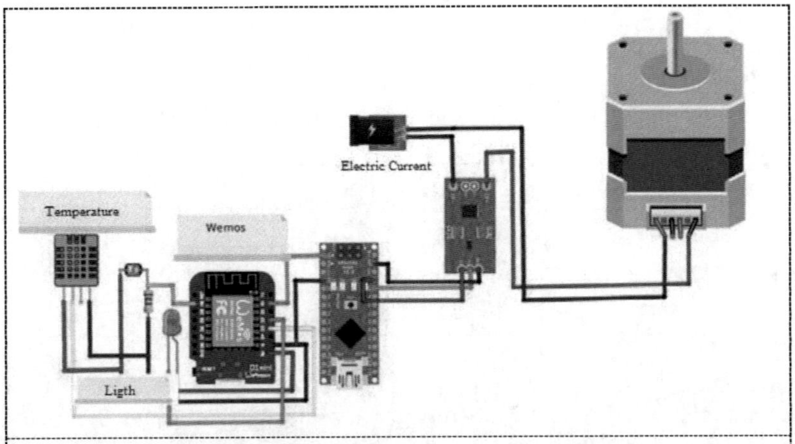

Figure 3. Schema hardware system monitoring refrigerator.

The hardware design that the authors do includes designing a block diagram of the prototype tool, and the electronic module used in making IoT-based refrigerator monitoring prototypes.

Variables are used to detect issues in the refrigerator.
The light conditions on the refrigerator lamp consist of three conditions: Normal (X1), Faint (X2) and Off (X3)

Table 2. The of Variable Range of refrigerator's light Condition

Variable	Range
Normal	More than 56 ohm
Faint	11-58 ohm
Off	0-10 ohm

The table above is sourced from research conducted with technicians at PT. Xyz, namely the variable conditions to determine the condition of the refrigerator lamp which consists of normal, dim and dead variables, for the variable value using Ohm units. The threshold value above is from the variable category and the value range is obtained from the technician at the research site.

The condition of the cooling temperature gauge in the refrigerator consists of three conditions Cold (X4), Less Cold (X5) and Off (X6).

Table 3. Table of Variable Range of Refrigerator's Temperature

Variable	Range
Cold	0-11 Celcius
Less Cold	9-21 Celcius
Off	More than 19 Celcius

The table above is sourced from research conducted with team technicians at PT. Xyz, namely the variables to determine the temperature conditions of the refrigerator consisting of cold, less cold and dead variables, for the variable value using Celsius units. The threshold value above is from the variable category and the value range is obtained from the technician at the research site.

The condition of the electric current meter in the refrigerator consists of three conditions Normal (X7), Not Normal (X8) and Off (X9).

Table 4. Table of Variable Range of Electric Current

Variable	Range
Normal	99-121 watt
Not Normal	More than > 191 watt
Off	Less than 99 watt

The table above is sourced from research conducted with technicians at PT. Xyz, namely the variable to determine the condition of the refrigerator electric current which consists of normal, abnormal and dead variables, for the variable value using watts. The threshold value above is from the variable category and the value range is obtained from the technician at the research site.

3.2. Manual Fuzzy Logic Calculation Process

The following is a manual calculation for the process of determining whether the refrigerator is damaged or not damaged. The first process is to create a matrix of refrigerator conditions, process rules, calculations and results.

If the condition is damaged and the value of cold temperature, normal light and normal current is given a value of 0 because the following conditions are not damaged and if the temperature is not cold enough, the temperature is off, the light is dim, the light is off, the current is not normal and the current is off then the value is 1 because this condition is the result is broken

If the condition is not damaged and the value of cold temperature, normal light and normal current is given a value of 1 because the following conditions are not damaged and if the temperature is not cold enough, the temperature is off, the light is dim, the light is off, the current is not normal and the current is off then the value is 0 because of this condition the result is damaged.

Table 5. Table of Rule Conditions

Number	Conditions
1.	IF SD – CN – AN THEN TR
2.	IF SD – CN – ATN THEN R
3.	IF SD – CM – AM THEN R
4.	IF SD – CR – AN THEN R
5.	IF SD – CR – ATN THEN R
6.	IF SD – CR – AM THEN R
7.	IF SD – CM – AN THEN R
8.	IF SD – CM – ATN THEN R
9.	IF SD – CM – AM THEN R
10.	IF SKD – CN – AN THEN R
11.	IF SKD – CN – ATN THEN R
12.	IF SKD – CN – AM THEN R
13.	IF SKD – CR – AN THEN R
14.	IF SKD – CR – ATN THEN R
15.	IF SKD – CR – AM THEN R
16.	IF SKD – CM – AN THEN R
17.	IF SKD – CM – ATN THEN R
18.	IF SKD – CM – AM THEN R
19.	IF SM – CN – AN THEN R

20.	IF SM – CN – ATN THEN R
21.	IF SM – CN – AM THEN R
22.	IF SM – CR – AN THEN R
23.	IF SM – CR – ATN THEN R
24.	IF SM – CR – AM THEN R
25.	IF SM – CM – AN THEN R
26.	IF SM – CM – ATN THEN R
27.	IF SM – CM – AM THEN R

The table above is the table determining the rules from the combinations in the matrix, namely from the following combinations there are 27 rules.
Information :

- SD is Cold Temperature.
- SKD is Less Cold Temperature.
- BC is Dead Temperature.
- CN is Normal Light.
- CR is Low Light.
- CM is Off Light.
- AN is the Dead Current.
- ATN is Abnormal Current.
- AM is the Dead Current.
- TR is Undamaged.
- R is Damage.

If the refrigerator temperature is 2 Celsius and the refrigerator light is 70 Ohm and the current is 100 Watt.

$$\mu[x] = \begin{cases} \mathbf{0}; & \boldsymbol{x \leq a \; or \; x \geq d} \\ \mathbf{1}; & \boldsymbol{b \leq x \leq c} \\ (\mathrm{d}-\mathrm{x})/(\mathrm{d}-\mathrm{c}) & \mathrm{c} \leq \mathrm{x} \leq \mathrm{d} \end{cases} \quad (1)$$

$$\mathrm{COG} = \frac{\int_a^b \mu A(x) x \, dx}{\int_a^b \mu A(x) \, dx} \quad (2)$$

Here are the minimum values for a broken rule:
0, 0.010989010989010999, 0.9, 0, 0.010989010989010999, 0, 0, 0.9, 0, 0, 0, 0, 0, 0, 0, 0, 0, 0, 0, 0, 0, 0, 0, 0, 0, 0, 0
Here are the minimum values for an undamaged rule:
0,6818181818181818
Take the max value from the list of values for the broken condition: 0.9
Take the max value from the list of values for the undamaged condition: 0.6818181818181818
Calculate with the method of finding the center of gravity, the formula is as follows

$$\mathrm{COG} = \frac{\int_a^b \mu A(x) x \, dx}{\int_a^b \mu A(x) \, dx} \quad (3)$$

$$\frac{(0.1+0.2+0.3+0.4+0.5) \times MaxA) + ((0.6+0.7+0.8+0.9+1) \times MaxB)}{(MaxA \times 5) + (MaxB \times 5)} \quad (4)$$

Calculation of the formula above:

Result formula: (0.1 + 0.2 + 0.3 + 0.4 + 0.5) * 0.9) + ((0.6 + 0.7 + 0.8 + 0.9 + 1) * 0.6818181818181818) / (0.9 * 5) + (0.6818181818181818 * 5)
Result formula = 4.077272727272726 / 7.99090909090908 = 0.5155172413793103

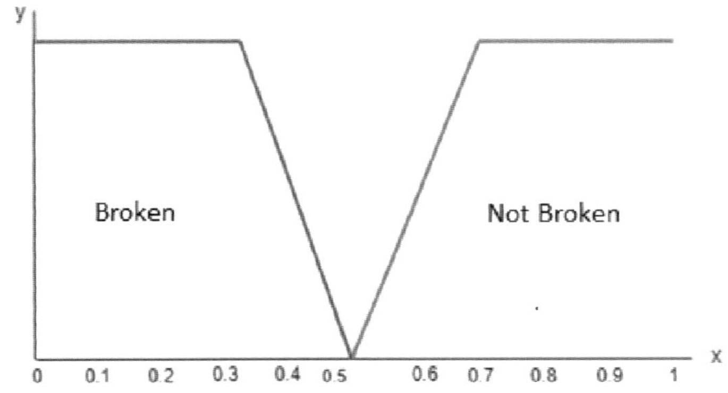

Figure 4. Fuzzy sets for broken and not broken variable.

Image of fuzzy set for damaged and undamaged variables, In the Figure above it can be seen that:
Value 0.1,0.2,0.3,0.4,0.5 including the set Damaged.
Values above> 0.5 include the Undamaged set.
From the above calculations, the value is 0.5155172413793103 if 0.5155172413793103 is entered into the graph, the result is the condition is not damaged

4. Conclusion

The use of the Fuzzy Logic Sugeno method in this research is expected to detect damage to the refrigerator from the data obtained from the temperature, light and current sensors. With the creation of this system, it is expected to be able to monitor the refrigerator accurately and quickly using IoT.

References

[1] I Wayan Degeng And Michael Santoso 2018 Smart Kulkas Dengan Fitur Sms Untuk Melaporkan Ketersediaan Bahan Makanan
[2] Deepti Singh And Preet Jain 2016 IoT Based Smart Refrigerator System
[3] Floarea, A.-D., & Sgarciu, V. 2016 Smart refrigerator: A next generation refrigerator connected to the IoT
[4] Fielding, R. T. and Taylor, R. N. 2002 Principled design of the modern Web architecture. Transactions on Internet Technology 115–150
[5] Jacquline W and Oktoverano L. 2018 Indoor Air Quality Monitoring and Notification System

Electronic Participation Technology Adoption Model in Indonesia: The Role of Gamification Impact

D Napitupulu[1,2], K Adiyarta[1], W Windihastuty[1] and T Fatimah[1]

[1]Faculty of Information Technology, Universitas Budi Luhur, Jl. Ciledug Raya, Jakarta Selatan, Indonesia
[2]Research Center for Science, Technology and Innovation Policy and Management, Indonesian Institute of Sciences, Banten, Indonesia

E-mail: darmawan.napitupulu@budiluhur.ac.id

Abstract. This study aims to propose a conceptual model in the context of e-participation technology adoption. This research develops the UTAUT base-line model with several constructs, namely Insecurity and Gamification. The impact of gamification to increase user participation is not widely known, especially regarding the acceptance of e-participation. The method used is a literature study related to several best practice models such as UTAU, TRI and gamification constructs. The results showed a total of 7 constructs and 6 hypotheses proposed in this study. All these constructs are performance performance expectancy, effort expectancy, social influence, facilitating conditions, insecurity, gamification impact and behavioral intention to use. The implication of this research is that the lack of public participation in digital platforms can be increased through redesign of e-participation by applying the gamification approach.

1. Introduction

In the last two decades, e-Government has been adopted by the government with the aim of providing public information and services online [1]. In addition, Presidential Instruction No. 3/2003 states that the public also wants their aspirations to be heard in the formulation of state policies. Therefore, the government must facilitate public participation starting from the process of making government policies to their implementation [2]. The lack of public participation will create resistance to government policies. Failures in development programs in the past were also caused by low public participation in the process of formulating, implementing and evaluating government programs [3]. As a result, development is no longer participatory and aspirational. The low public participation also causes the poor quality of public services provided by the government. The rampant practice of extortion, convoluted bureaucratic processes, non-transparent procedures show low government integrity [4]. This will reduce the level of public trust in the government. According to [5], the benefit of public participation in policy formulation is to gain legitimacy in the form of public trust in the government. When this legitimacy is obtained, the implementation of government policies will be easier because there is no resistance from the community. Therefore, public participation is an important factor in the formulation of state policies where the role of society at this time is not only as recipients but also contributing (co-create) in producing public policies and services [6].

Along with these conditions, research with a focus on e-Government to encourage public participation in government has also increased [7] - [9]. Information and Communication Technology

IOP Conf. Series: Materials Science and Engineering 1071 (2021) 012031 doi:10.1088/1757-899X/1071/1/012031

(ICT) is used as a tool to facilitate public participation, defined as e-participation. E-Participation is more specifically the use of ICT to conduct online consultation and public dialogue [10]. E-Participation comes in various forms ranging from simple online applications to interactive electronic platforms such as social media, online discussion forums and mobile applications [11]. All of these channels aim to provide the widest possible opportunity for the public to participate in conveying aspirations, information and even criticism as a form of evaluation of government policies.

Unfortunately, the lack of public participation is still a challenging issue for the current government [11]. It can be said that the e-Participation that has been developed so far has not been widely used by the public. E-participation initiatives generally experience problems when participation is done online. Some of the factors that cause it, among others, the public is not familiar, low public interest, the design is less attractive to its complexity [10], [12]. This is in line with several studies that still highlight the lack of public participation in e-participation [8], [11]. Thus, further study is needed related to the factors of e-participation adoption in order to encourage public participation in government administration.

The purpose of this research is to develop a gamification-based e-participation model to increase public participation. Gamification is a game-based approach that is still relatively new in the government domain, aimed at providing a positive experience that can motivate users to participate actively [13]. Research related to the impact of gamification on e-participation is still very limited [14]. Therefore the contribution of this research is to fill the existing literature by proposing an e-participation model that combines the UTAUT model as a baseline theory with the gamification construct..

2. Methodology

Research related to technology adoption models has actually been carried out by many previous researchers where various behavioral theories emerged related to user behavior in adopting a technology. Popular behavioral theories are Theory of Reasonable Action (TRA), Theory of Planned Behavior (TPB), Theory of Diffusion of Innovation (DIT), Technology Acceptance Model (TAM), Technology Acceptance Model 2 (TAM 2), Technology Readiness Index (TRI), Delon & McLean (2003) and Unified Theory of Technology Acceptance and Use (UTAUT) [13]. The background for the emergence of behavior theory is due to the rapid and dynamic development of technology itself. This of course has an impact on how fast users can accept and adopt a technology, depending on a number of factors such as convenience, usability, convenience, speed, security, etc. [15]. Each behavior theory in the field of technology, especially information technology (IT) proposes different technology adoption factors from different perspectives.

Behavioral theory can be used to help organizations predict whether a technology can be adopted successfully or not by its users. Attempts to understand why people accept or reject a technology is very challenging [16]. It is also important to know so that the technology that has been developed can be used by its users in the future. In addition, behavior theory aims to find out in depth the user's behavior and attitudes by asking users direct questions to find out their perceptions about the use of certain systems / technologies. Then the user will evaluate the technology according to their experience using the technology. Behavioral theory is a branch of psychology that studies human behavior scientifically where response behavior can be observed and measured. In other words, user behavior can be studied and explained scientifically. Behavior is a response to stimuli from the environment that can be learned.

In this study, the base-line behavior theory used to predict user behavior in adopting e-participation is UTAUT (Unified Theory of Acceptance and Use of Technology). The UTAUT theory or model was found by Venkatesh (2003) to explain user behavior towards technology, especially information technology or information systems (IT / SI) [17]. The UTAUT theory or model shows that behavioral intention and behavior using a technology are influenced by performance expectation, effort expectation, social influence and facilitating conditions. These four factors are moderated by factors of gender, age, experience and voluntary use. Many empirical studies that adopt this model have been conducted and have various findings. However, these four factors, namely performance expectation, effort expectation, social influence and facilitating conditions, have been widely used to predict user acceptance and adoption of technology [18,19,20].

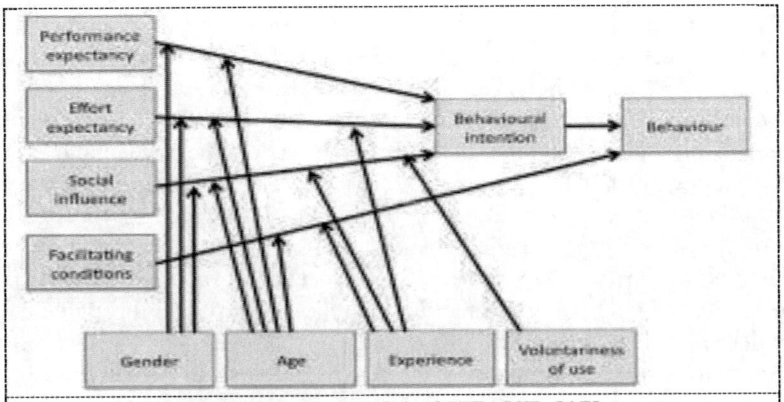

Figure 1. Model of UTAUT [17]

The UTAUT model is a combination of eight previously developed models, namely Theory of Reasoned Action (TRA), Technology Adoption Model (TAM), Motivational Model (MM), Theory of Planned Behavior (TPB), Combined TAM and TPB (C-TAM)). -TPB), PC Use Model (MPCU), Diffusion Innovation Theory (IDT), and Social Cognitive Theory (SCT). The UTAUT model proved to be more successful than the other eight theories in explaining up to 70 percent of the user variant [17]. After evaluating these eight models, Venkatesh (2003) found six constructs that appear to be significant direct determinants of behavioral intention or usage behavior in one or more of each model. These constructs are performance expectation, effort expectation, social influence and facilitating conditions, attitude toward use of technology and self-efficacy. After conducting further testing, it was found that four main constructs play an important role as direct determinants of behavioral intention and usage behavior, namely, performance expectation, effort expectation, social influence and facilitating conditions. Others proved insignificant as direct determinants of behavioral intention. In addition there are also four moderators: gender, age, experience and voluntary use which are positioned to moderate the impact of the four main constructs on behavioral intentions and user behavior. Figure 1 shows the UTAUT model and the relationship between its constructs. The main purpose of research using UTAUT is to help organizations understand how users react to a new technology [21]. Initially, UTAUT was developed from the Technology Acceptance Model (TAM) in 2003 in an organizational context with four main constructs. However, currently UTAUT has been developed from an organizational context into an individual context, namely UTAUT 2 where habit factors, hedonic motivation, and price values are added as new constructs in the model..

In addition, in this study, the UTAUT model will be combined with several constructs such as insecurity and gamification in the context of adopting e-participation. The insecurity construct is one of the constructs of the TRI (Technology Readiness Index) model developed by Parasuraman (2000) which is used to measure user readiness in adopting a technology [17]. Insecurity can be interpreted as insecurity or suspicion of a technology regarding personal data. Insecurity also refers to distrust of technology-based transactions as well as doubts about the workability of a technology. Meanwhile, gamification is widely applied to increase user participation in adopting technology. Gamification is believed to be able to influence, involve, and motivate individuals, groups or communities so as to encourage the creation of behavior and produce the desired effects and results [22]. At the same time, gamification can reduce perceived barriers to system use such as low usability or difficulty to use where minimal public participation remains an issue the government has not resolved to date.

3. Result and Discussion

The UTAUT (Unified Theory of Acceptance and Use of Technology) model is a theory or model that has been widely used to analyze the factors that influence user acceptance or adoption of technology. In other words, UTAUT provides information related to individual behavior toward technology that can be measured from the user's perception. UTAUT is a development of the previously popular theory, namely TAM (Technology Acceptance Model) by adding two external constructs, namely Social Influence and Facilitating Conditions. This is because the weakness of the TAM model is that it only focuses on internal factors, namely the technological aspects. Furthermore, UTAUT 2 produced several external constructs (outside the technological aspect) which proved significant, such as hedonic motivation, habit and price value [18-20]. However, on the other hand, UTAUT still has a weakness, namely it does not consider the psychological factors of users which are also proven to influence users' decisions in adopting a technology [17]. Therefore, in this study, the proposed construct is one of the user's psychological factors derived from the TRI (Technology Readiness Index) model, namely Insecurity. The insecurity construct in the context of this study is related to one's belief in the ability of a technology to facilitate public participation through electronic platforms. Insecurity is considered an obstacle because of the distrust of activities or transactions that occur in the electronic platform, in this case e-participation. Users will feel insecure when their personal identity is known by others in submitting information / complaints / aspirations to the government through e-participation. In addition, the gamification construct was also added to the UTAUT model in the context of this study. As with the problem related to the lack of public participation in digital platforms, e-participation needs to be redesigned with the aim of increasing community involvement to participate actively through the gamification approach [23, 24].

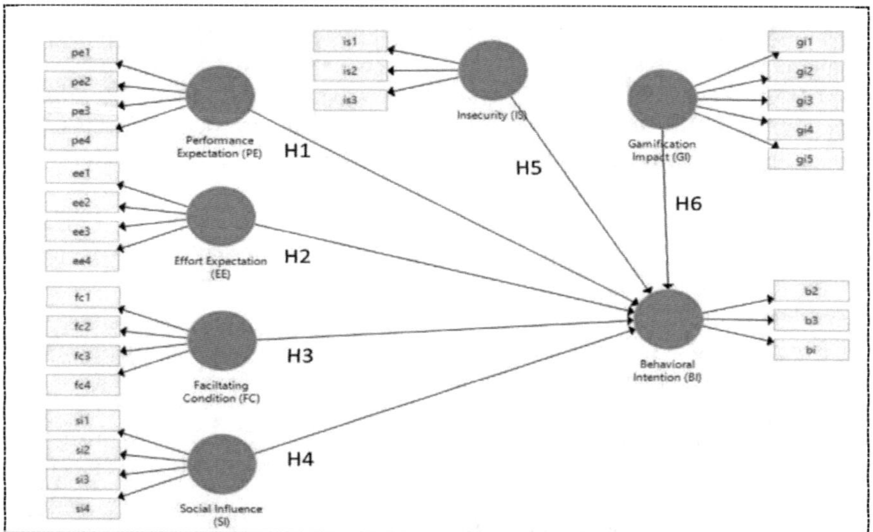

Figure 2. Proposed Research Model

Gamification is the use of game mechanics and game design techniques in a non-game context to design behavior, develop skills, or to engage people in innovation [25]. Gamification is a new, effective way to increase motivation or involvement of people to solve complex problems, take certain actions, or just have fun [26]. Although gamification has been applied to various domains, the impact or influence of gamification on technology adoption, especially e-participation, is still very limited. Research conducted by Baptista & Oliveira (2017) shows the potential impact of using mechanics and game design techniques (gamification) on user acceptance in the financial industry [27]. Unfortunately the research was conducted in a different context from this research, namely on mobile banking services. Therefore, the gamification construct is added in this study because it is believed that applying game techniques in a non-game context such as e-participation will have a significant impact on increasing the level of user acceptance, as shown in Figure 2.

Based on Figure 2, there are 6 hypotheses generated from the proposed conceptual model which can be described as follows:

H1: Performance Expectation (PE) has a significant positive effect on Behavioral Intention (BI)
H2: Effort Expectation (EE) has a significant positive effect on Behavioral Intention (BI)
H3: Social Influence (SI) has a significant positive effect on Behavioral Intention (BI)
H4: Facilitating Condition (FC) has a significant positive effect on Behavioral Intention (BI)
H5: Insecurity (IS) has a significant negative effect on Behavioral Intention (BI)
H6: Gamification Impact (GI) has a significant positive effect on Behavioral Intention (BI)

The combination of the UTAUT (The unified theory of accept and use of technology) construct with the gamification construct is used as a theoretical support model for investigations in this study according to Figure 2. UTAUT has been empirically tested and proven to be superior to other models that exist today [28]. The inclusion of the gamification construct in the proposed research model allows researchers to gain a better understanding of the impact of this gamification factor, which is believed to be one of the important driving factors for the level of acceptance of e-participation in the future. Other constructs as moderating variables in the UTAUT model such as gender, age, experience and voluntary use were not considered in this study.

4. Conclusion

The results of this study propose a conceptual model of electronic participatory adoption (e-participation) by combining the UTAUT model with several constructs such as Insecurity and the Gamification impact construct. Research related to the impact or potential of gamification on user acceptance is still limited in the context of e-participation. The contribution of this research is to fill in the lack of literature related to how gamification affects user participation in digital platforms. In addition, the proposed model also includes user psychological factors in the form of insecurity to determine the level of user security in using e-participation. In the context of government, users should be given a sense of security in submitting complaints or aspirations to the public services provided. All 6 hypotheses in this study require further empirical research to test proving the proposed e-participation adoption model.

Acknowledgement

The author would like to thank for the funding of Hibah Terapan Scheme Contract Number 225/SP2H/LT/DRPM/2019 from Direktorat Riset dan Pengabdian Masyarakat, Direktorat Jenderal Penguatan Riset dan Pengembangan, Kementerian Riset, Teknologi, dan Pendidikan Tinggi Republik Indonesia in order to conduct the research successfully.

References

[1] M. K. Feeney and A. Brown, "Are small cities online? Content, ranking, and variation of U.S. municipal websites," Gov. Inf. Q., vol. 34, no. 1, pp. 62–74, 2017.

[2] D. Napitupulu, "User-Acceptance instrument development: a content validity study in the e-participation context," J. Appl. Res. Technol., vol. 18, pp. 34–43, 2020.

[3] I. W. A. Suarnata, A. T. Atmaja, N. Luh, and G. Erni, "Pengelolaan Alokasi Dana Desa (Studi Kasus Pada Desa Manikliyu Kecamatan Kintamani Kabupaten Bangli)," S1 Ak Univ. Pendidik. Ganesha, vol. 8, no. 2, 2017.

[4] H. Puspitosari, Khalikussabir, and L. J. Kurniawan, Filosofi Pelayanan Publik: Buramnya Wajah Pelayanan Menuju Perubahan Paradigma Pelayanan Publik. Setara Press (Kelompok Intrans Publishing) dan Jaringan Nasional [dengan] Masyarakat Peduli Pelayanan Publik (MP3), 2011.

[5] B. R. Pflughoeft and I. E. Schneider, "Social media as E-participation: Can a multiple hierarchy stratification perspective predict public interest?," Gov. Inf. Q., vol. 37, no. 1, 2020.
[6] J. Ju, L. Liu, and Y. Feng, "Public and private value in citizen participation in E-governance: Evidence from a government-sponsored green commuting platform," Gov. Inf. Q., vol. 36, no. 4, 2019.
[7] X. H. Wang and T. A. Bryer, "Assessing the Costs of Public Participation: A Case Study of Two Online Participation Mechanisms," Am. Rev. Public Adm., vol. 43, no. 2, pp. 179–199, 2013.
[8] J. Lee and S. Kim, "Citizens' e-participation on agenda setting in local governance: Do individual social capital and e-participation management matter?," Public Manag. Rev., vol. 20, no. 6, pp. 873–895, 2018.
[9] Y. Zheng, H. L. Schachter, and M. Holzer, "The impact of government form on e-participation: A study of New Jersey municipalities," Gov. Inf. Q., vol. 31, no. 4, pp. 653–659, 2014.
[10] M. Toots, "Why E-participation systems fail: The case of Estonia's Osale.ee," Gov. Inf. Q., vol. 36, no. 3, pp. 546–559, 2019.
[11] J. Choi and C. Song, "Factors explaining why some citizens engage in E -participation , while others do not," Gov. Inf. Q., vol. 37, no. July, 2020.
[12] L. Hassan and J. Hamari, "Gameful civic engagement: A review of the literature on gamification of e-participation," Gov. Inf. Q., vol. 37, no. 3, 2020
[13] R. N. Landers, E. M. Auer, A. B. Collmus, and M. B. Armstrong, "Gamification science, its history and future: Definitions and a research agenda," Simul. Gaming, vol. 49, no. 3, pp. 315–337, 2018.
[14] S.-K. Thiel, T. Petra Ertiö, and M. Baldauf, "Why so serious? The role of gamification on motivation and engagement in e-participation," Interact. Des. Archit. J., vol. 35, no. December 2017, pp. 158–181, 2017
[15] Silva PM, Dias GA. Theories about technology acceptance: why the users accept or reject the information technology? Brazilian J Inf Sci [Internet]. 2007;1(2):69–86
[16] Davis FD, Bagozzi RP, Warshaw PR. User Acceptance of Computer Technology: A Comparison of Two Theoretical Models. Manage Sci [Internet]. 1989;35(8):982–1003. Available from: http://pubsonline.informs.org/doi/abs/10.1287/mnsc.35.8.982
[17] Venkatesh V, Morris M, Davis G. User Acceptance of Information Technology: Toward a Unified View. MIS Q. 2004;28(4):695–704
[18] Zuiderwijk A, Janssen M, Dwivedi YK. Acceptance and Use Predictors of Open Data Technologies: Drawing upon The Unified Theory of Acceptance and Use of Technology. Government Information Quarterly. 2015; 32:429-440
[19] Cimperman M, Brencic MM, Trkman P. Analyzing older users' home telehealth services acceptancebehavior—applying an Extended UTAUT model. International Journal of Medical Informatics. 2016; 90: 22-31
[20] Isaac O, Abdullah Z, Aldholay AH, Ameen AA. Antecedents and outcomes of internet usage within organisations in Yemen: An extension of the Unified Theory of Acceptance and Use of Technology (UTAUT) model. Asia Pacific Management Review. 2019; 24: 335-354
[21] Al-Gahtani SS, Hubona GS, Wang J. Information technology (IT) in Saudi Arabia: Culture and the acceptance and use of IT. Inf Manag. 2007;44(8):681–91.
[22] Rodrigues, L. F., Costa, C. J., & Oliveira, A. How gamification can influence the web design and the customer to use the e-banking systems. In Proceedings of the International Conference on Information Systems and Design of Communication - ISDOC '14 (pp. 35–44). 2014. New York, USA: ACM Press. doi:10.1145/2618168.2618174
[23] Morschheuser B, Hassan L, Werder K, Hamari J. How to design gamification? A method for engineering gamified software. Inf Softw Technol. 2018;95:219–37.

[24] Huotari K, Hamari J. A definition for gamification: anchoring gamification in the service marketing literature. Electron Mark. 2017;27(1):21–31.
[25] Burke, B. Gamification 2020: What is the future of gamification? Gartner. 2012. Retrieved from https://www.gartner.com/doc/2226015/gamification--future-gamification
[26] Mishra, R. K. Infosys Labs briefings gamification: Rediscover the power of engagement. Infosys Labs Briefings, 2013; 11(3)
[27] Baptista G, Oliveira T. Why so serious? Gamification impact in the acceptance of mobile banking services. Internet Research. 2017; 27(1)
[28] Park, J., Yang, S., & Lehto, X. Adoption of mobile technologies for chinese consumers. Journal of Electronic Commerce Research. 2007; 8:196–206

Identification of Factors Affecting Electronic Participation: a Literature Review

Titin Fatimah[1], Krisna Adiyarta[1], Darmawan Napitupulu[1,2] and Martini[3]

[1]Faculty of Information Technology, Universitas Budi Luhur, Jl. Ciledug Raya, Jakarta Selatan, Indonesia
[2]Research Center for Science, Technology and Innovation Policy and Management, Indonesian Institute of Sciences, Jakarta, Indonesia
[3]Faculty of Economics and Business , Universitas Budi Luhur, Jakarta, Indonesia

E-mail: titin.fatimah@budiluhur.ac.id

Abstract. This study aims to conduct a systematic review of technology aadption factors of electronic participation applicaton in Indonesia. Application of LAPOR as Online People's Aspirations and Complaints Service has been developed to fulfill the public demands to actively participate in the formulation of government policies. However, there is limited understanding about technology acceptance regarding to users' behaviour especially in Indonesia. The method used is a systematic review with the PRISMA approach, consisting of five stages, namely defining eligibility criteria; define the source of information; literature selection; data collection and item selection with synthesis techniques. The results showed that there were 40 constructs of technology adoption of LAPOR application. The implication of this research is that government institutions could better understand what important factors that influence the adoption of technology.

1. Introduction

Based on Presidential Instruction No. 3 of 2003 concerning policies and strategies for developing e-Government, there are two public demands that must be fulfilled by the government, namely quality public services and public participation[1]. Regarding the second claim, the government must facilitate public participation and dialogue in the formulation of state policies. This is because basically the masses want their aspirations to be heard in the context of participating in overseeing the ongoing and ongoing government programs and policies.

In the opinion of Cohen & Uphoff (1977), one form of public participation is when people can express their opinions to assess, criticize, ask or provide input on a plan or program that will be determined and on a decision or policy that is running[2]. Through e-Government, people can now easily convey their aspirations to the government. One form of convenience from the benefits provided by e-Government is a simpler procedure in channeling aspirations without having to go directly to the government office. The community now only needs to access services through the government website provided. This certainly affects the shorter bureaucratic pathways, cost savings and time needed by the community. Thus, e-Government can be said to increase public participation where participation is carried out in the ICT domain or called e-Participation. Phang & Kankanhalli (2008) also explains that e-Participation occurs when citizens are empowered to actively participate by utilizing ICT in an online environment[3,4]. In addition, ICT also allows the public to participate in several ways such as chat as a media or discussion forum, profiles that allow users to be part of a system and information provision that allows the public to convey information.

LAPOR application is one form of embodiment of e-Participation that serves as a service to handle the aspirations or complaints of citizens as seen in Figure 1. This application was developed by UKP4 (Presidential Work Unit for Development Monitoring and Control) and managed jointly with the Ministry of State Apparatus and Bureaucratic Reform (KemenPANRB). The application aims to provide convenience for the public to provide aspirations or complaints. So far, many residents have found it difficult to report complaints. In fact, not a few people also feel that they have not received a response to the complaints raised, or their complaints are often thrown here and there so that citizens feel disappointed with such treatment. Even though this has been regulated in Law No. 25 of 2009 concerning public services, in article 18 it is stated that citizens have the right to get responses to complaints submitted. More than that, the community also has the right to notify the government to improve its services if it is not yet in line with service standards. Facts on the ground, residents do not know how and where to go if they want to make a complaint[5].

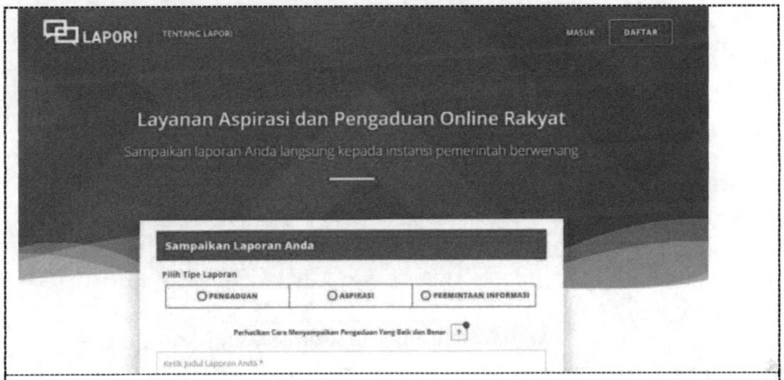

Figure 1. LAPOR Website for Public Participation [9]

During the first administration of Susilo Bambang Yudhoyono (SBY), he made a breakthrough in communicating SMS (Short Message Service) and PO BOX mailboxes. The community can directly submit their complaint via a complaint and address SMS or PO BOX with number 9949. The complaint number is flooded with 500-600 SMS every day and within the first 3 months has received 13,045 complaints. The letters were immediately returned, some were channeled to the relevant departments and regional heads to be verified. The contents of letters and SMS vary greatly from extortion cases, illegal logging, corruption to personal problems. This method is actually quite effective in capturing citizen input and his efforts to get closer to people from all over Indonesia. But unfortunately the effectiveness of this program was not evaluated and the results were not reported to the public. Citizens cannot help the government's performance in completing their complaints reports. Residents also do not know the existence of the report they provided, whether the report has been followed up or not [6,7].This is because the program cannot display all incoming public complaints and follow-up from related agencies or institutions. Another disadvantage of the program is that there is no certainty of time for reporters.

To overcome these weaknesses, UKP4 took the initiative to build a LAPOR Application which is supported by a communication channel, the LAPOR website! (www.lapor.go.id), mobile apps, social media Twitter (@ LAPOR1708) and SMS number 1708[7]. LAPOR application is one form of implementation of the mandate of Law Number 25 of 2009 concerning Public Services specifically Presidential Regulation Number 76 of 2013 concerning Management of Public Service Complaints. LAPOR application as a National Public Service Complaint Management System, it can be used as a forum for people to interact with the government more easily, while being able to actively participate in overseeing development and public services. At the same time, the government has the means to capture community participation and map existing problems while finding solutions[5].

According to Mahendra et. al (2017), there are three main advantages to using the LAPOR application namely first, LAPOR an integrated information system has connected around 22,217 government agencies both ministries, institutions and local governments that are spread throughout the country so that the government can more easily collaborate with the public in overseeing development and public services. Second, the LAPOR application wide opportunities for the government to increase transparency and accountability because citizens are now able to track the whereabouts of their reports which are always documented and published on the website. Third, LAPOR empower citizens to provide reports more easily because without going through a complicated bureaucratic process, by utilizing various website technologies, mobile applications, social media and SMS [7,8]. It is proven that until 2019, the LAPOR apps has received as many as 605,847 reports from the public addressed to 208 ministries, 173 institutions, 847 regional agencies, 261 Regional Work Units (SKPD) and 2435 other agencies through various LAPOR channels [9].

Based on the benefits and facilities provided by the LAPOR application it further facilitates citizens to submit complaints reports regarding government programs and policies and public services provided. Until April 2015, LAPOR users the number has increased significantly to 290,000 with an average of more than 800 public reports per day[9]. LAPOR application also succeeded in encouraging the government to be more responsive and able to increase the closeness of public relations with the government. This is evident that the government has become more responsive to reports, aspirations and information conveyed by citizens. The 2014 data shows the government's response was able to resolve up to 78% of the total reports of residents who entered. The remaining reports are still in the process of being resolved and or have not been followed up. In other words, the LAPOR application able to eliminate bureaucratic divides between government and society [7].

Referring to the above indications, it can be said the LAPOR application said to be quite successful in becoming a nationally integrated media of public aspirations and complaints. This is also supported by various awards obtained such as winning the top three nominations for Government web award at Bubu award 2013 and one of the world's best initiatives at the 2013 Open Government Partnership Summit in London[9]. Therefore, in this study a review was made regarding the determinants of user acceptance from the LAPOR application. This is because research on evaluating citizen participation in the national scope in Indonesia is very limited, especially the LAPOR application. Thus the theoretical contribution of this research will complement the knowledge gap in the form of literature on citizen participation in the application of e-Government in Indonesia. Practically, this research can provide recommendations for the government in making policies related to the implementation of other e-Government systems, especially to increase citizen participation.

2. Method

Research on evaluating e-participation in the government nationally in Indonesia is very limited as a result of the lessons learned about what is a prerequisite and knowledge of the determinants of user acceptance of e-participation can also be said to be minimal. This study aims to conduct a systematic review of previous studies related to the factors of user acceptance of e-participation, especially the LAPOR application. The method used in this study is Qualitative Systematic Review with the PRISMA approach (Preferred Reporting Items for Systematic Reviews and Meta-Analyzes).

PRISMA is a reporting technique that is used to present reports optimally from systematic review activities so that the results can be utilized effectively. PRISMA was discovered by Liberati et. al (2009) which has advantages in terms of transparency and complete reporting [10]. The stages of PRISMA can be presented in Figure 2 as follows:

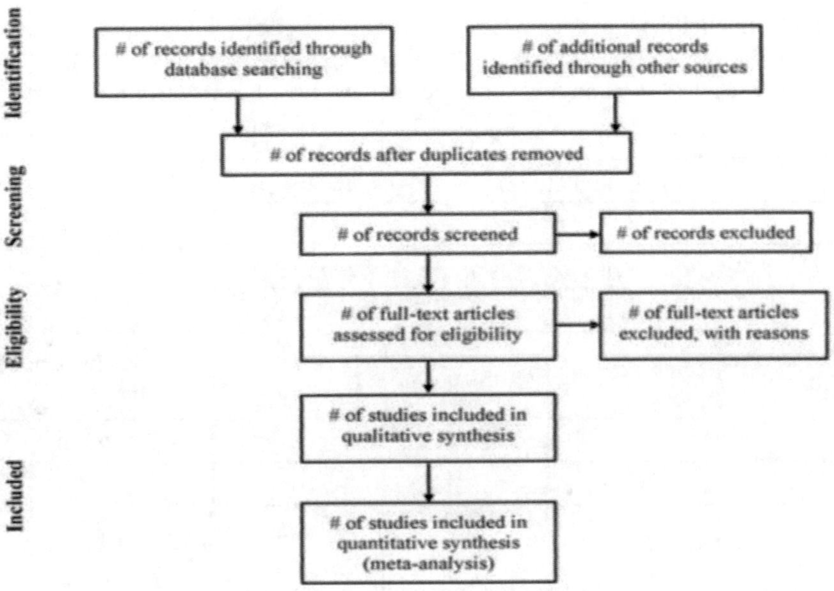

Figure 2. PRISMA Phase Diagram [10]

In the PRISMA approach, the first important thing to do is to determine the eligibility criteria. This will determine the quality of the results of the systematic review. In this study, there were 3 eligibility criteria as follows:

IC-1: Refereed Research written in English and Indonesian Language related to the study of e-Participation especially the LAPOR application namely the national public service complaint management system. The implication is that there are quite a lot of research papers from Indonesian researchers.

IC-2: Research paper published in the past five years because it is relevant to the emergence of the LAPOR application since 2013-2014.

IC-3: Research is aimed at investigating citizen's acceptance factors with the LAPOR application in Indonesia.

In addition to the eligibility criteria, information sources are also an important element in the search for the desired article. In this study, information sources were accessed from Google Scholar and IEEE electronic databases. This is due to the limitations of obtaining studies related to citizen's acceptance factors from the LAPOR application through other reputable databases like Science Direct and Elsevier (Scopus). The process of searching for articles is done based on keywords that have been determined, namely "e-participation" AND "LAPOR". After the article was obtained, further screening of articles that were significantly relevant to the research theme, namely citizen's acceptance factors of LAPOR as shown in Figure 1. The process of article selection is done in layers to get quality articles through screening duplications, titles, abstracts, keywords, body of text, references and eligibility based on predetermined eligibility criteria. Furthermore, the process of data collection and selection of data items were carried out as the last step to report the results of the systematic review of citizen's acceptance factors from the LAPOR application.

IOP Conf. Series: Materials Science and Engineering 1071 (2021) 012032 doi:10.1088/1757-899X/1071/1/012032

3. Result and Discussion

Based on the stages in PRISMA discussed in the previous section, the authors identified studies and screening based on eligibility criteria (IC-1 till IC-3) to obtain studies relevant to the objectives of this study. The selection process (screening) in this research can be presented in Figure 2 as follows:

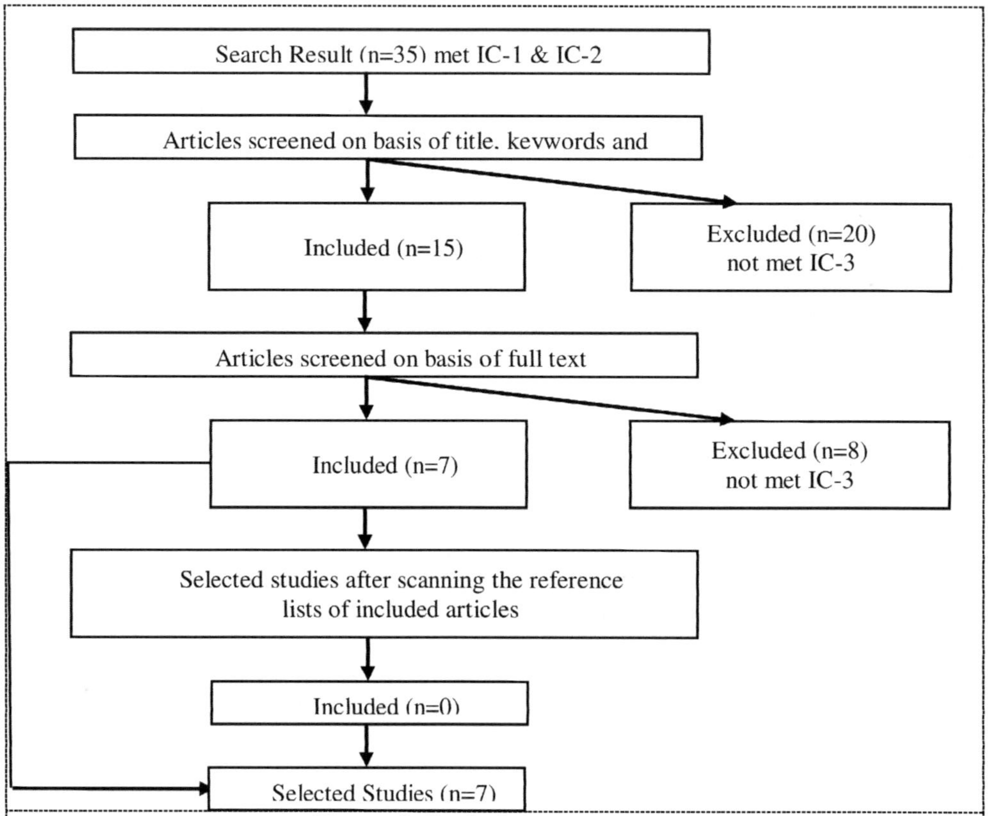

Figure 3. Study Selection Process.

Based on Figure 3, it can be seen that the results of the screening (screening) in the database initially produced 35 articles according to keywords and were written in English (IC-1) and published in the last five years 2014-2019 (IC-2). Furthermore, the articles were selected based on titles, abstracts and keywords, leaving only 15 relevant articles. 20 articles were excluded from the list because they did not meet the IC-3 eligibility criteria. Then the entire article is re-selected based on on base of full text. The total articles involved in the next process were 7 articles that were significantly relevant to the research theme where 8 articles were excluded because they did not meet the IC-3 criteria and there were no additional articles produced from the reference list.

The next stage is collecting data from 7 cross studies consisting of database source information, author, article title, publication year as presented in Table 1 below:

Table 1. Summary of Selected Study

No	Database	Author	Title	Publication Year
1.	IEEE	Khoirunnida et. al [11]	Factors influencing citizen's intention to participate in e-participation: integrating Technology Readiness on Social Cognitive Theory	2017
2.	Google Scholar	Rido Panjaitan[8]	Government in online spaces: critical evaluation of citizen-to-government participation in urban centres in java, indonesia	2018
3.	Google Scholar	Fathul Wahid[12]	Affordances and effects of promoting eParticipation through social media	2015
4.	Google Scholar	Dinur Sadat[13]	M-government implementation evaluation in encouraging citizen participation in indonesia: a case study of LAPOR	2014
5.	Google Scholar	Wahyu Mahendra, et. al[7]	Citizens' Aspirations and Complaints Online System (LAPOR) in Indonesia: Making Citizens Happy	2014
6.	Google Scholar	Nataya Anindita[5]	Elemen Sukses E – Government: Studi Kasus Layanan Aspirasi Dan Pengaduan Online Rakyat (LAPOR) Kota Bandung	2016
7.	Google Scholar	Rheina Wulansari[14]	Peningkatan partisipasi warga negara dalam pelayanan publik melalui pemanfaatan laman layanan aspirasi dan pengaduan online rakyat (LAPOR) Di kota bandung	2019

Table 1 above shows the results of collecting studies that are relevant to the research. In total there are 7 cross-study articles related to citizen acceptance factors from e-Participation, especially the LAPOR application. Articles obtained are in accordance with the IC-1 eligibility criteria (written in English and Indonesian) and most articles come from within the country. There are only 1 (one) article published in IEEE International Proceedings. This is because the theme of the research raised is related to the LAPOR national-based application. The entire article also meets the IC-2 eligibility criteria because of the range of article publications from 2014-2019 and according to the IC-3 criteria that are relevant to the investigation of citizen's acceptance factors from the LAPOR application.

Furthermore, the next phase of PRISMA and is the last phase, namely the process of selecting data. Synthesis techniques are used to integrate all factors related to citizen acceptance of the LAPOR application in Indonesia. The results of data selection can be shown in Table 2 below:

IOP Conf. Series: Materials Science and Engineering 1071 (2021) 012032 doi:10.1088/1757-899X/1071/1/012032

Table 2. Synthesis of LAPOR Acceptance Factors

No	Citizen's Acceptance Factors	Relevant Study						
		1	2	3	4	5	6	7
1.	Ease of use & Simple Process	V		V			V	V
2.	Cutting-of Bureaucracy Process	V					V	V
3.	Interactive & Communicative Services	V						
4.	Clarity and certainty of service	V						
5.	Service innovation and Multi-Features	V	V	V			V	
6.	Perceived Cost and Time of Citizen	V						
7.	Routinely Monitoring	V						
8.	Vision& Mision	V						
9.	Supportive Policy	V						
10.	Reward & Punishment	V			V			
11.	Leader's knowledge and personality	V						
12.	Transparaceny & Accountability	V			V			
13.	Cost & Time Efficiency of Government	V						
14.	Increased involvement of citizens	V						
15.	Collaboration	V						V
16.	Citizen knowledge and concern	V						V
17.	Socialization	V					V	V
18.	Motivation	V						V
19.	Outcome Expectation Personal		V					
20.	Participation Self Efficacy		V					
21.	Optimism		V					
22.	Insecurity		V			V	V	
23.	Convenience							
24.	Integrated System& Multi-Channel	V		V			V	V
25.	Compatibility			V			V	
26.	Legal Framework			V				V
27.	Top Management Commitment/Political Will			V	V			
28.	Budget		V	V			V	
29.	Adequate Technical Skill Staff				V		V	V
30.	Focus IT Shifting				V			
31.	Social Media Use				V			
32.	Support from parliament				V			V
33.	System Quality	V	V	V		V	V	V
34.	Information Quality					V		V
35.	Service Quality					V		
36.	User Satisfaction					V		
37.	Intention to Continual Use					V		
38.	Trust of Citizen					V		
39.	Cross-Border Service						V	
40.	Fast follow-up/Responsiveness	V	V	V		V		

Table 2 above shows that the results of data selection with synthesis techniques produce a total of 40 factors that influence citizen acceptance of factors towards the e-Participation program, especially the LAPOR application. The synthesis process is carried out on 7 cross-studies that are significantly related to the research theme, namely citizen's acceptance factors of LAPOR. The entire citizen 's acceptance factors are of an important nature that should be applied by government institutions to be able to increase citizen participation in the future through the application of e-Participation especially the LAPOR application in Indonesia.

4. Conclusion

This study has successfully demonstrated the systematic stage of literature review with the PRISMA (Preferred Reporting Items for Systematic Reviews and Meta-Analyzes) approach and reports the results of synthesis related to citizen acceptance factors for the application of the LAPOR in Indonesia. The contribution of research theoretically enriches the literature on citizen's acceptance factors towards e-Participation especially the LAPOR application which has been applied to the national scope since 2013. Research on evaluating citizen participation in the national scope in Indonesia is still very limited and therefore the contribution of this research to fill the knowledge gap has been carried out. The results of the study show that 40 citizen 's acceptance factors were obtained from both internal (government) and external (citizen) sides. All of these factors should be applied by both central and regional government agencies that want to implement the LAPOR application. The expected output is an increase in citizen participation, especially in terms of supervision of development and public services through the LAPOR application.

Acknowledgement

The author would like to thank for the funding support of Direktorat Riset dan Pengabdian Masyarakat, Direktorat Jenderal Penguatan Riset dan Pengembangan Kementerian Riset, Teknologi, dan Pendidikan Tinggi Republik Indonesia so the research could be conducted successfully.

References

[1] Instruksi Presiden Nomor 3. Kebijakan dan Strategi Nasional Pengembangan E-Government. 2003.
[2] Uphoff N, Cohen MJ. Rural development participation: Concepts and measures for project design, implementation, and evaluation. Ithaca, N.Y.: Rural Development Committee, Center for International Studies, Cornell University. 1977. 317 p.
[3] Slaviero C, Cristina A, Garcia B, Maciel C. Electronic Participation. 2012;8654(September). Available from: http://link.springer.com/10.1007/978-3-662-44914-1
[4] Phang CW, Kankanhalli A. A framework of ICT exploitation for e-participation initiatives. Commun ACM [Internet]. 2008;51(12):128. Available from: http://portal.acm.org/citation.cfm?doid=1409360.1409385
[5] Anindita N. Elemen Sukses E – Government: Studi Kasus Layanan Aspirasi Dan Pengaduan Online Rakyat (LAPOR) Kota Bandung. 2016.
[6] Detiknews. Sebelum Era Twitter , SBY Terima Aduan Masyarakat lewat SMS dan PO BOX. 2014.
[7] Mahendra W, Pratiwi M, Prawesti R. Citizens' Aspirations and Complaints Online System (LAPOR) in Indonesia: Making Citizens Happy. In: WCPA. 2014. p. 1–15.
[8] Panjaitan RP. Government in Online Spaces: Critical Evaluation Of Citizen-To-Government Participation In Urban Centres In Java, Indonesia. 2018.
[9] Kantor Staf Kepresidenan. Statistik LAPOR 2019.
[10] Liberati A, Altman DG, Tetzlaff J, Mulrow C, Gøtzsche PC, Ioannidis JPA, et al. The PRISMA statement for reporting systematic reviews and meta-analyses of studies that evaluate health care interventions: explanation and elaboration. J Clin Epidemiol [Internet]. 2009;62(10):e1-34. Available from: http://www.ncbi.nlm.nih.gov/pubmed/19631507
[11] Khoirunnida, Hidayanto AN, Purwandari B, Yuliansyah R, Kosandi M. Factors influencing citizen's intention to participate in e-participation: Integrating Technology Readiness on Social Cognitive Theory. Proc 2nd Int Conf Informatics Comput ICIC 2017. 2018;2018–January:1–7.
[12] Wahid F, Saebo O. Affordances and Effects of Promoting eParticipation Through Social Media. In: 7th International Conference on Electronic Participation (ePart). 2015. p. 3–14.
[13] Sadat DR. M-Government Implementation Evaluation in Encouraging Citizen Participation in Indonesia: A Case Study of LAPOR [Internet]. 2014. Available from:

/citations?view_op=view_citation&continue=/scholar%3Fhl%3Des%26start%3D130%26as_sdt%3D0,5%26scilib%3D1%26scioq%3Dtools&citilm=1&citation_for_view=PfIHHdUAAAAJ:4n0clTBhZ78C&hl=es&oi=p

[14] Wulansari R. Peningkatan Partisipasi Warga Negara Dalam Pelayanan Publik Melalui Pemanfaatan Laman Layanan Aspirasi Dan Pengaduan Online Rakyat (LAPOR) Di Kota Bandung. 2019.

AUTHOR INDEX

Institute of Physics
Dirac House, Temple Back
Bristol BS1 6BE UK

ISSN: 1757-8981
ISBN 978-1-7138-3345-1
9 781713 833451

CURRAN ASSOCIATES INC.
proceedings
.com